PRAISE FOR
The Lab Rat Chronicles

"Witty, wise, and down-to-earth, Kelly Lambert teaches us much about ourselves by letting her rats do the talking. By understanding the fundamentals of brain function in rats, we can see reflections of human nature. Lambert is a spellbinding storyteller, and in finest form, she tells us her favorite brain stories, provoking new insight into ourselves."

—Patricia Churchland, Professor Emerita, Philosophy Department, University of California–San Diego, and author of *Braintrust*

"Lambert's book does for the brain and animal behavior what Julia Child did for cooking. She takes the arcane, hard-to-digest material one usually finds in academic neuroscience journals and renders it into a delectable concoction. Funny, clever, and engaging, Lambert shows that doing good science is both important and fun."

—Craig Kinsley, PhD, Professor of Neuroscience, Department of Psychology, University of Richmond, and coauthor of *Clinical Neuroscience*

"In this very interesting book, Kelly Lambert does a terrific job of explaining to a nonscientific audience what deep matters can be explored by clever experiments on a by-no-means-simple animal."

—Peter Sterling, PhD, Professor of Neuroscience, University of Pennsylvania School of Medicine

"A best-kept secret of brain science is that laboratory rats and mice have told us as much, perhaps more, about our basic nature than the study of humans. Kelly Lambert gracefully shares the saga of how the study of these fascinating, friendly, and curious animals has illuminated the psychological complexity of rat nature that can now inform us about our own minds and habits."

—Jaak Panksepp, Baily Endowed Professor of Animal Well-Being Science, College of Veterinary Medicine, Washington State University, and author of *The Archaeology of Mind*

"Kelly Lambert elevates the rat, the selfsame creature who bedevils subways and dark alleys, to our indispensable doppelgänger. Linking her experience in the trenches of rat research with the issues that challenge her own family, Kelly Lambert illustrates how *Rattus norvegicus* has enlightened us on everything from work to play, courtship, child rearing, and addiction. If man ever gets a second-best friend, it should be the rat."

—Hannah Holmes, author of *The Well-Dressed Ape* and *Quirk*

The Lab Rat Chronicles

A Neuroscientist Reveals
Life Lessons from the Planet's
Most Successful Mammals

KELLY LAMBERT, PhD

A Perigee Book

A PERIGEE BOOK
Published by the Penguin Group
Penguin Group (USA) Inc.
375 Hudson Street, New York, New York 10014, USA
Penguin Group (Canada), 90 Eglinton Avenue East, Suite 700, Toronto, Ontario M4P 2Y3, Canada
(a division of Pearson Penguin Canada Inc.)
Penguin Books Ltd., 80 Strand, London WC2R 0RL, England
Penguin Group Ireland, 25 St. Stephen's Green, Dublin 2, Ireland (a division of Penguin Books Ltd.)
Penguin Group (Australia), 250 Camberwell Road, Camberwell, Victoria 3124, Australia
(a division of Pearson Australia Group Pty. Ltd.)
Penguin Books India Pvt. Ltd., 11 Community Centre, Panchsheel Park, New Delhi—110 017, India
Penguin Group (NZ), 67 Apollo Drive, Rosedale, Auckland 0632, New Zealand
(a division of Pearson New Zealand Ltd.)
Penguin Books (South Africa) (Pty.) Ltd., 24 Sturdee Avenue, Rosebank, Johannesburg 2196,
South Africa

Penguin Books Ltd., Registered Offices: 80 Strand, London WC2R 0RL, England

While the author has made every effort to provide accurate telephone numbers and Internet addresses at the time of publication, neither the publisher nor the author assumes any responsibility for errors or for changes that occur after publication. Further, the publisher does not have any control over and does not assume any responsibility for author or third-party websites or their content.

First edition: June 2011

Library of Congress Cataloging-in-Publication Data

Lambert, Kelly.
  The lab rat chronicles : a neuroscientist reveals life lessons from the planet's most successful mammals / Kelly Lambert.
    p. cm.
  "A Perigee book."
  Includes bibliographical references and index.
  ISBN 978-0-399-53663-2
  1. Rats—Behavior. 2. Rats—Psychology. 3. Rats as laboratory animals. 4. Psychology, Comparative. I. Title.
  QL737.R666L33   2011
  599.35'2—dc22   2011004991

PRINTED IN THE UNITED STATES OF AMERICA

10 9 8 7 6 5 4 3 2 1

Most Perigee books are available at special quantity discounts for bulk purchases for sales promotions, premiums, fund-raising, or educational use. Special books, or book excerpts, can also be created to fit specific needs. For details, write: Special Markets, Penguin Group (USA) Inc., 375 Hudson Street, New York, New York 10014.

For my students—past, present, and future
and, of course . . . the lab rats

ACKNOWLEDGMENTS

||||||||||||||||||||||||||||||||||||||||||||||||||||||||||||||||||||||||||

As the title of this book implies, the bulk of the material was provided by my investigations of the resourceful and adaptive rats that have occupied my Randolph-Macon laboratory for the past two decades. I am eternally grateful to my students who have contributed a continuous string of innovative and informative research with the rodents through the years. Many of these students are named in the book but, due to space and editing limitations, many more are not; regardless, my heartfelt thanks is extended to each and every student who has worked in some capacity in my laboratory. A special thanks is also extended to Catherine Franssen, a former student who has returned to R-MC as a postdoctoral fellow—her presence has certainly enhanced the productivity of the lab. The research described in *The Lab Rat Chronicles* has been supported by several R-MC funds generously provided by the following sources: Chenery Research Grants, Walter William Craigie Grants, Schapiro Undergraduate Research Fellowships, the Macon and Joan Brock Professorship, the Duff family of Fredericksburg, and the Randolph-Macon Psychology Department. Additionally, valuable support has been provided by the National Institutes of Health and the National Science Foundation.

Beyond the laboratory, several individuals contributed to the emergence of this book. A special thanks is extended to my agent, Michelle Tessler, and Perigee editor, Marian Lizzi—both of whom saw value in a rather unique book proposal emphasizing the life lessons that could be gleaned from the world of rodents. I am also indebted to valued colleagues in the R-MC Psychology Department

ACKNOWLEDGMENTS

as well as a supportive provost (William Franz) and president (Robert Lindgren). Barb Wirth has made immeasurable contributions by providing excellent administrative assistance throughout this process. Outside of R-MC, colleagues around the world have generated valuable research providing essential content for this book. My most frequent research collaborator, Craig Kinsley, has contributed to several research stories chronicled in this book; hopefully the next couple of decades yield equally interesting research adventures. Massimo Bardi, a more recent collaborator, has infused valuable insights into our rodent work with his expertise in behavioral and endocrinological analyses. I am also indebted to several colleagues who enhanced the quality of the manuscript by generously providing valuable feedback prior to the final edits of the book. Any mistakes or misinterpretations in the final manuscript, however, are my sole responsibility.

Because I maintained my "day job" as a professor and researcher while writing this book, my nights and weekends were frequently occupied with writing. My family remained supportive throughout this busy time and always listened enthusiastically as I described new rat discoveries for various chapters. My husband, Gary, went above and beyond spousal editing obligations by providing insightful comments and edits for each and every chapter. As mentioned in the book, my daughters, Lara and Skylar, have restructured my neural networks in ways that have enhanced the meaningfulness of life's many endeavors. I appreciate their love and tolerance of a rather unconventional mom who is a self-declared rodent enthusiast!

CONTENTS

⁗⁗

||||||||||||||||||||||||||||||||||||||||||||||||||||||||||||||||||||

CHAPTER 1

Whisker Wisdom
Emerges

Can rodents reveal the central secrets of the success of mammals? Can they illuminate the complex inner mental lives of humans, the most advanced species on the planet? Indeed, the idea of scientists working alongside rats to try to solve the many mysteries of the mind seems more like a plot from a Disney animated movie than a nonfiction scientific tale. This unlikely scenario, however, is my professional reality and places me in the rather unique position of being able to tell the rats' amazing story of survival. Typical of most unauthorized biographies, secrets will be revealed; in this case, secrets of the rodents' adaptive life strategies that may benefit other opportunistic species, especially humans.

My career as a behavioral neuroscientist over the past quarter of a century has been defined by my ability to work effectively and learn from my rodent colleagues. Designing experiments that allow rodents to "speak their mind" in their own unique ways is the crux of my scientific career. The consequences are clear—the rodents don't cooperate, I don't have data to evaluate my latest hypothesis. Without such data, there are no manuscripts to submit for publication, no talks to give at conferences, no informative studies to share with my undergraduate students, and no pilot data for grant proposals. Such consequences would lead to my professional demise.

I realize that I'm in a minority—few people have valuable professional relationships with rodents and some folks may even find the idea of rats disgusting, thinking that they pose nothing but a threat to our well-being. My hope is, via some informative research and personal stories and insights, to move beyond this unfair species profiling. As Beatrix Potter's books suggested over a century ago, we can learn a lot about ourselves by observing the lives of rodents (and other animals). As you'll learn throughout this book, the scientific tales are so informative and interesting that there's no longer a need for a fictional story line. Without the cute clothes and make-believe tea parties characteristic of Potter's books, there are many rich life lessons to learn from rodents. In fact, after reading this book I hope you'll think about the rodents when you face your next challenge in life, perhaps wondering, What would the rodents do? I'm getting ahead of myself; before regaling you with their impressive accomplishments, an introduction is in order.

A Little Rat History

The notion of placing rodents on a laboratory or literary pedestal has certainly not always characterized the relationship between humans and rodents. Rats immigrated to the United States by stowing away on ships traveling from various areas of Europe. Being both opportunists and omnivores, humans and rats have a lot in common. Rats were clever enough to determine that wherever humans were, suitable food would be available, and a beautiful relationship emerged. Hopping off the ships to the cities was an easy adjustment for these adaptable rodents, and they were just as successful at colonizing the United States as were their human counterparts. Known as a *commensal* mammal because it shares our table by extracting food from humans without causing too much of a disturbance, this relationship took a new turn about a century ago when scientists saw an opportunity to hire rats for their research investigations.

Before their journey to the New World, however, rats fell on hard times in their European homeland. A popular sport in France and England around 1800 was rat baiting. This violent sport involved placing a trained terrier in a pit with one to two hundred rats and taking bets on how long it would take until all the rats were killed. In the 1830s, rat fighting also became a favorite pastime in the United States—as anxious immigrants entered New York City, they would reportedly relax by cramming into little saloons and watching the rats fight. Considering the tragedies rats (and their fleas) brought to humans in the forms of deadly plagues before this time, this violent treatment was somewhat expected. Few saw value in these animals. But beauty is in the eye of the beholder. The massive breeding necessary to produce enough of these animals for the baiting contests also produced an occasional albino rat, an exotic and prized animal, even if it was a rat.

Through the years, the albino rats were kept as pets—and became tame and docile. As these pet rats were bred, they became increasingly compliant and easy to handle. In fact, it was these rare albino versions of the common brown rat that would eventually change research as we know it today. It is thought that Swiss psychiatrist Adolph Meyer, one of the most influential psychiatrists of the first part of the twentieth century, brought the first albino rats to the United States when he immigrated in 1890 to join the faculty at the University of Chicago. It was there that he met Henry Herbert Donaldson, who would eventually leave his appointment in the Department of Neurology to become the research director of the Wistar Institute in Pennsylvania. Under Donaldson's leadership, the first official commercial rat colony was established. Painstaking research revealed important features of these animals that paved the way for them to transform biomedical research. Today, 85 percent of all biomedical research in the world is conducted on rats and mice.

Two of the most popular contemporary rat varieties are the albino rat, complete with its white coat and red eyes and, my personal favorite, the Long-Evans hooded rat, with its black head (looks like a hood) and stripe going down its otherwise white fur. Although both strains appear very different from their wild brown rat ancestor (also known as *Rattus norvegicus* because there was some confusion about them originating from Norway—an observation later found to be untrue as we now know they originated in Asia), these laboratory strains are still viewed as card-carrying members of the brown rat family. It is estimated that one scientific publication based on the responses of laboratory rats is produced every hour. Other species of rats also exist but have not become popular career rats in the laboratory. Because they are smaller and more economical to house, mice are also frequently employed in labs across the world. Although we can learn a lot from the mice, I remain most impressed with the rat. It has been written that the rat chose complexity whereas the mouse chose simplicity as a survival trait. Consequently, the rat has slowly morphed into a social, skilled, intelligent, and complex species.

It is estimated that one scientific publication based on the responses of laboratory rats is produced every hour.

The transformation of interactions between rats and humans is quite an American success story. In fact, it was recently stated that the laboratory rat story is literally one of ascendancy from the gutter to a place of nobility, for what creature is "more lowly than the rat as a wild pest or more noble than the same species that has contributed so much to the advancement of knowledge as the laboratory rat?"

As scientists domesticated rats and began various types of research with these animals, they muddled along with their studies because they considered laboratory rats and mice to be the perfect generic laboratory mammals. Many of the scientific projects seemed easy: Inject substance X into generic mammal and record effect. Although the initial research on rats was critical for establishing scientific success in the laboratory, little consideration was given to the fact that this recently domesticated mammal was the sophisticated product of millions of years of evolution. Consequently, it would be beneficial to understand more about the rat's natural history to understand the reams of data being collected every day in laboratories across the world. As behavioral neuroscientists became increasingly familiar with laboratory rodents, it became more and more obvious that there was much more to these mammals than their reflexive physiological and behavioral responses. Our success as scientists was going to require us to get to know these mammals, and their wild counterparts, on a personal level.

A Crash Course from the Wild Ones

I've spent a lot of time in the laboratory observing and examining rats, but I have to admit that one of the most valuable lessons learned from a rat was not in one of my carefully designed research studies, nor was it gleaned from one of the nearly one hundred neuroscience conferences, chock-full of rodent studies, I have attended over the past several decades. The strongest dose of *whisker wisdom* I have ever encountered presented itself a few years ago when a wild rat made an unexpected visit to my office in the Copley Science Center.

> A wild rat made an unexpected visit to my office.

The fact that a rat would make its way from the wild to the office and lab of a behavioral neuroscientist who spent her career investigating the rat brain and rat behavior was beyond ironic. This animal broke into my office one morning, ripping open a ketchup packet in my desk drawer and leaving its indisputable rodent paw prints all over my desk. As I examined the rest of my office and adjacent lab for further evidence of my rodent roommate I was both horrified and excited to learn that this rat had built a nest on one of the hidden shelves a mere eight feet from my desk. And what did this resourceful animal use for building materials? I laughed after realizing that the nest was made of strips of paper from a University of Richmond alumni magazine that featured my col-

I had encountered an *academic* rat.

league Craig Kinsley and one of his albino laboratory rats on the cover. The second building material was even more creative—bits of sheep brains that had been used for the brain practical in my behavioral neuroscience class. It was obvious that I was not dealing with a typical city, country, or even laboratory rat . . . I had encountered an *academic* rat.

College policy required me to call a professional pest control technician to capture this wild pest before further damage could be done. Although the professional claimed that this task would be easy, each attempt to trap the academic rat was met with failure. Various types of live traps were used, and each time the rat managed to get the "goodie" without activating the trap. From the beginning, I have to admit that I was rooting for the rat, thrilled each time it outsmarted the professional rat catcher.

After weeks of observing the rat's ability to outsmart the traps designed by the supposedly more superior human brain,

I was sitting in my office one quiet Memorial Day and heard something fall from a shelf in my lab. After turning around, I locked eyes with my rodent office mate for the first time. He had found the sugar and powdered milk I use for some of my histological protocols and knocked the bags to the ground when I startled him. I managed to enclose the rat under a bookshelf but, even with the help of a lab assistant—and buckets, and gloves, and nets—I was not able to catch this resourceful animal. I had him in my net momentarily, but his strength and agility were no match for us, and he soon escaped. This was a dramatic contrast to my laboratory rats, which I could handle like kittens with no apparent objections. Although I wanted to examine the brain of this clever animal, I secretly cheered as he ran across my feet, implementing his exit strategy. As I jumped around doing the laboratory version of the Riverdance each time the rat ran by, I knew my fourteen-hundred-gram brain had been outsmarted by an animal with a mere two grams of neural tissue! I was more intrigued than ever before.

What lesson did I learn from my wild rat visitor? When given the opportunity to solve problems outside of the protected confines of the laboratory, rats exhibit impressive intellect and resourcefulness. This experience was humbling and reminded me to respect the natural habitats and lifestyles of these animals as I attempt to learn more about them in the laboratory. If an alien research team ventured down from outer space and tried to learn more about my behavior by observing me in a boring wire cage, they wouldn't learn very much. To my regret, most scientists are forced to observe their rodent subjects in the laboratory as opposed to a natural habitat. As described later, however, the new trend in scientific circles is to create more natural environments for our animals—even when the rats are assessed in the laboratory. If we want to learn more about the true nature of our own mental lives, we have to respect the true nature of the rodent's lives we spend so much time studying. This approach is our only chance of creating adequate models of behavior that may reveal secrets about the many challenges facing our species, such as depression, autism, cardiovascular disease, and immune diseases.

Model Behavior

What is an animal model? Far from walking the runway in couture fashion shows, the top animal models offer our best hope for discovering information that will lead to cures for the various diseases that haunt us. Let's not forget that animals were used to test and perfect the once far-fetched, almost science-fiction idea of organ transplants. Other animal models have been used to understand the effects of a plethora of drugs on medical conditions ranging from high blood pressure to diabetes. Although we are still far from a cure, animal models have led to many of the existing treatments for cancer. We

> The top animal models offer our best hope for discovering information that will lead to cures for the various diseases that haunt us.

can't mimic a human's complex physiology by limiting our investigations to cells in a Petri dish; we need a representative animal model to test these life-changing hypotheses. In most cases, scientists have turned to rats and mice as a first step in answering these research questions.

A challenge for biomedical scientists using rodent models is the double-edged sword produced by the sterile environment required to maintain the rodents' health and the efficiency of traditional laboratory investigations. In many cases, an unfortunate consequence of the rigid regulations found in labs across the world is the fact that these environments have become so artificial that the laboratory rat may be compromised as a representative mammalian model, especially as a model of various aspects of mental health and mental illness (which is my area of interest).

Along these lines, I often read about criticism of the human fashion model industry, stating that the models are so tall and thin that they don't represent *real* women. When I was a teenager growing up in rural Alabama, I remember anxiously awaiting the arrival of my *Seventeen* magazine each month—a window to a world outside of my small hometown. Each month, however, I became increasingly disappointed as I realized that the models in the magazines, with their stick-like bodies and bizarre clothing styles, could never represent a teen in my world. I recall writing a letter to the magazine's editor stating that the publishers should include clothing models representative of all teen girls across the nation; further, I mentioned that if I dressed like their models—with a weird hat, bizarre boots, distinct eye makeup, and wild-colored leggings—I would more likely be admitted to a mental institution than impress anyone with my fashion sense. I received a reply thanking me for my views and asking me to consider doing an internship with them when I got a little older; so much for trying to change the fashion world!

A similar argument can be made for animals that are housed or tested in artificial laboratory environments. As I compare the behavior of my laboratory animals to the memory of the James Bond–like behavior of the wild rat that visited my lab, I constantly question the validity of our rodent work in the lab. Accordingly, a high priority in my lab and other behavioral neuroscientists' labs across the world is the consideration of species-typical natural behaviors of these resilient animals. Even with the limitation of laboratory research, rodents have allowed us to learn massive amounts of information underlying the complexities of behavior. With a new respect for their natural behaviors, dedicated scientists will unveil increasingly important secrets of mammalian success.

Now that so much is known about the natural history of rats, I cringe when I view old footage of the earliest behavioral studies involving these animals. Rats were frequently housed individually in cold metal cages. The floors of the cages were made of metal grids so the feces and urine could pass to a lower collection pan, allowing no solid floor for the rats to build a nest, which is so integral to the existence of their wild rodent cousins. Further, experimental tests involved artificial variables such as shock, restraint, and bar pressing.

As students enter my laboratory today, they see vast differences between the typical experimental scenarios in the 1940s and the current research strategies. Animals are housed in the company of other animals to enable rich social interactions.

They have solid floors and different types of bedding materials to build nests. Various novel objects are placed in their cages to enhance their cognitive development. The temperature, lighting, and humidity are monitored closely to reduce their stress and maximize their health. Contemporary laboratory rodents receive only healthy food for consumption (unlike many of today's children); however, a few sweet treats or seeds are given to animals as rewards in various learning tests. In short, we want to create a healthy and somewhat natural and complex environment to ensure that our laboratory animals are representative models of their wild counterparts. There are inevitable differences between laboratory and wild rats (discussed later in the book), but we are satisfied that we are doing the best we can to generate realistic behaviors in laboratory-bred animals.

What valuable information may be gleaned from the use of more realistic rodent models in the laboratory? An increased vigilance about relevant animal models is critical for the development of adequate animal models of mental illness. Is it possible to generate a single perfect animal model of depression? My answer is no, but it is possible to capture key symptoms of depression in various animal models, allowing scientists to create an informed picture of what is going on during such emotional forays. For example, we have found that animals exposed to chronic stress decrease their motivation to drink sugar water, a treat for laboratory rats. This evidence suggests that their pleasure response is altered by chronic stress, an observation to which we all may be able to relate. This model also provides an opportunity for exploring the underlying brain mechanisms accompanying diminishing pleasurable responses, known as *anhedonia* in the neuroscience literature. Focusing on the responses of rats in this type of rodent depression model may provide critical clues about brain activity that characterize depression in humans suffering from this disorder.

In my laboratory, my students and I have also found connections between physical activity and a decreased tendency to give up on life's problems. How did we determine this in rodents? Rats given the opportunity to work for their rewards (the worker group) were more persistent when they encountered a subsequent unsolvable problem than were rats that were merely given their rewards, regardless of effort (the trust fund rats). Instead of giving up immediately, the animals that have made strong connections between their behavioral effort and positive consequences appear to be more resilient and resist the temptation to give up. We'll discuss this in further detail in Chapter 3. Again, we can explore the underlying neurobiological components of depression as we explore a set of behavioral systems and accompanying brain circuits that represent the cluster of symptoms we recognize as depression in humans.

A Shared Biology Leads to New Insights

Although the behaviors described in the chapters that follow are the products of millions of years of evolutionary tweaking, you'll be surprised at just how timely rodent lessons can be for humans, the evolutionary new kids on the block. The recent unveiling of the rat genome project confirmed that we possess about 90 percent of the same genes. As the rodent lineage split to form the rat and mouse lineages somewhere between twelve and twenty-four million years ago, it appears that they adapted to changing demands of the environment with more genetic refining than found in primate lineages eventually leading to modern humans. In fact, the rat is the highest achiever when it comes to accelerated evolutionary adaptations, meaning it can respond to changes in the environment faster than other mammals. The enhanced adaptation rate led to enhanced survival rates in changing and new environmental terrains.

Talk about success stories, the order Rodentia is the largest group of mammals, consisting of nearly seventeen hundred species inhabiting virtually every terrestrial habitat on the planet. The rats, mice, and their rodent relatives represent 40 percent of all mammalian species, making them, in my mind, one of the most successful groups of mammals in the world. Further, the genus *Rattus*, with its approximately fifty-five species, is the largest mammalian genus. Francis Collins, who formerly led the National Center for Human Genome Research and now directs the National Institutes of Health, conveyed his respect for rodent research by claiming they are "among nature's most resourceful and adaptive creatures and one of our close relatives."

The shared biology between humans and rodents is essential to the premise of this book. Although my political preferences don't always align themselves with conservative perspectives, my scientific preferences are exclusively conservative. This scientific partisan approach simply means that I recognize Nature's use of efficient mechanisms in the grand scheme of species development. Essentially, basic fundamental processes were conserved as new species emerged. How inefficient would it be for all mammals to have different types of brain cells, circulatory systems, immune cells, or completely specific movement systems? New adaptations certainly accompanied each new species, but with more commonalities with its closest relatives than differences. A young upstart executive who leaves her job with a Fortune 500 company to branch out on her own will most certainly not forget everything learned in the previous work environment. Certain practices will be tweaked and modified here and there, but the fundamentals will likely remain the same.

A recent experience reinforced the notion of conservation among mammals. One of my students was investigating how quickly rats with varying coping strategies recovered from an immune challenge, a small dose of a bacterium in this case that made the rats feel nauseous. As I walked into the lab to do a quick check on the animals' conditions I noticed several rats displaying a very distinct response. Their mouths were opening in a gaping pattern; they were moving their heads back and forth and moving their forelimbs in a manner suggesting that they were wiping something off of their faces. I immediately recognized this response as a "disgust/dislike" response, no different from my own response when I accidentally take a big sip of soured milk. In fact, University of Michigan neuroscientist Kent Berridge has conducted a systematic comparative analysis between humans and rats showing very similar responses toward aversive tastes as well as appetitive or pleasing tastes. The neural programming for these behavioral responses appears to be very similar in rats and humans.

What about more complex emotional responses? Neuroscientist Jaak Panksepp and his colleagues have provided compelling evidence that rats . . . well . . . they laugh! How do you know when rats are laughing? They emit a vocalization, or a chirp, in the fifty-kilohertz range when they are playing with other rats (a fascinating behavior discussed in Chapter 6), as well as when they are being "tickled" by an experimenter's hand. My students were amazed when they saw this response with their own eyes when Panksepp visited my campus several years ago and showed a video of one of his rat laughing/chirping sessions. After placing his hand passively in the rat cage, the rats immediately ran over to interact with it. When the hand suddenly started wiggling its fingers in a tickling motion, their chirping (played back in a range audible to human ears) was evident. When they approached the tickling hand, they chirped like children laughing in a bout of play wrestling with a parent. Of course, rat laughter is a difficult sell at somber neuroscience conferences; consequently, Panksepp and his colleagues have conducted painstakingly meticulous research to defend their claims. Their research has revealed that rat laughter chirps are positively correlated with the intensity of play. Further, young rats laugh more than older rats, and these laughter chirps are also observed when rats experience other forms of appetitive stimuli such as a much anticipated meal or electrical brain stimulation. A recent manuscript in the neuroscience journal *Behavioral Brain Research* titled "Neuroevolutionary Sources of Laughter and Social Joy: Modeling Primal Human Laughter in Laboratory Rats" confirms that laughter has arrived as a legitimate research area in seriously minded behavioral neuroscientists.

Another important aspect of rodents makes them invaluable models to use in search of the truth about human nature.

As I tell my students, the rats don't have a cover story. When I assess a rodent I get the real deal—pure unadulterated raw behavior. My rodents never tell me that they know the material but don't test well, or that they are visual learners (or in the case of the rats, per-

As I tell my students, the rats don't have a cover story.

haps I should say olfactory learners), or that they have test anxiety and need to be tested in an individual testing room. When I'm working with rats, I ask the question and I get the answer; of course, then I have the huge task of figuring out what that answer means. Although the human prefrontal cortex is revered in evolutionary circles, its amazing ability to fabricate stories to relieve our anxieties about various unexplained or unpredictable events in our lives is a substantial stumbling block for scientists conducting research on human subjects. One of my favorite examples of the human brain's tendency to distort reality is a research project that consisted of asking moms to report the aspects of their lives that brought them the most happiness. As scientists tabulated the results, participants indicated that a clear contributor to their personal happiness was their children; however, when asked to rate how happy their day-to-day activities made them, childcare rated right up there with housecleaning chores. Hmmm . . . which set of self-report data generated from the same brains do the scientists believe?

And even when expectations may not exist to thwart the analysis of our own behavior, our brains just aren't that good at reporting accurate information about what it has been doing.

John de Castro's work with college students' self-reports of the basic behavior of eating confirms this observation. When Georgia State University students were asked if they ate more in isolation or in social situations, they reported eating less when they were in the presence of other people. When asked to write down what they ate after each meal as well as the number of people present at the meal, the students' preliminary subjective reports of their consumption rates, compared to their actual eating diaries, were grossly inaccurate. Not only did food consumption not decrease in social gatherings, it increased—students ate about 45 percent more in social situations than when they were alone. In large groups, the meal size increased by 75 percent! As we evaluate the value of eyewitness testimony providing accurate information about other people's behavior, we need to take to heart these data suggesting that we can't even trust eyewitness testimony of our own lives. Obviously, there are other factors present when the frontal lobes spin their yarns to construct the mental fabric, or stories, of our lives. Many of these biases and inaccurate strategies exist for adaptive reasons, as described by cognitive psychologist Keith Stanovich, but this provides little comfort to scientists relying on human subjects to reveal the central truths of our mental lives.

Being knowledgeable of the challenges facing human researchers, the value of adding relevant animal models to the scientific literature becomes increasingly appropriate. Reviewing the reams of scientific findings in behavioral neuroscience laboratories across the world as well as drawing from my own rodent encounters over the past quarter of a century, I describe countless gems of knowledge throughout this book that can instruct us about seemingly uniquely human topics such as universal healthcare, conflict resolution, romance, intelligence, financial portfolios, and family values. Yes, feel free to be skeptical at this point. As I emphasize to my students, claims about natural truths—such as the ones I will be making throughout this book—demand empirical, or scientific, support. I encourage you to be critical and demand relevant findings to support any claims made in the forthcoming chapters. I pledge to provide this support and to try to keep my own frontal cortex from weaving interesting, but likely false, explanations of these fundamental questions about the nature of our day-to-day living.

A few years ago I was teaching a course about rat behavior, and when I began the discussion of the complex, yet predictable, grooming patterns in rats, I reflexively started waving my hands across my nose in increasingly larger concentric patterns as I reflexively modeled the grooming patterns of the rats. When I got to the middle part of my head and paused where my rodent ear should be, I looked at the concerned faces of my students as they were trying to decide if I was an expert in rat behavior or someone off her medication. Realizing that I could effortlessly groom myself in the same manner demonstrated by my rodent colleagues, I enthusiastically reported that indeed I had found my inner rodent!

After spending a year watching the rats of New York City in their urban habitats, journalist Robert Sullivan also seemed to have found a bit of his inner rodent. At the end of his book *Rats*, he wrote:

> As I was walking back to my apartment, I began to think more than ever that we are all a little like rats. We come

and go. We are beaten down but we come back again. We live in colonies and we strike out on our own, or get forced out or starved out or are eaten up by our competition, by the biggest rats. We thrive in unlikely places, and devour.

My hope is that, after reading this book, you'll search for your own inner rodent as you navigate through life's many challenges. Just as rodents test the waters by whisking their environment with their amazingly sensitive and informative vibrissae, or whiskers, before moving forward in a dangerous terrain, we can learn to use our own evolved senses and responses to make more informed decisions about taking the next step in the journey of life. Indeed, these rodent insights may be just the medicine—a *Rodent Rx* of sorts—that our species needs to keep it from losing touch with our ancestral environment. But I'm putting the tail of this tale ahead of the whiskers; there's a lot of territory to cover before we get to the most important lesson of all in the final chapter.

CHAPTER 2

||||||||||||||||||||||||||||||||||||||||||||||||||||||||||||||||||||||||||||||

Building the Brain Trust

I know what you're thinking. It's one thing to accept that we
share basic similarities with rodents—cellular mechanisms,
motor reflexes, an attraction to sweets—but the human brain
and its consequent mental abilities, that's something completely
different. It's fine to use rodent models to learn about the
effects of stress hormones on thymus gland weights, but leave
neuronal networks, attention spans, and metacognition to the
humans. Good point. I'm certainly not going to suggest that
rats harbor religious beliefs, align themselves with political
parties, or possess the cognitive capacity to strategize about
future military tactics—good or bad, those are uniquely human
endeavors. But the brain's contributions to our sense of personal awareness and ability to process information in meaningful and relevant ways didn't jump on the mental scene when
humans evolved. In this chapter we'll explore what the rodents

can reveal about the superior mental capacities often thought to be uniquely human.

Anthropo . . . What?

When I was in graduate school, over a century after Darwin's suggestion of a continuum between human and nonhuman mammals, my professors drilled in me the importance of keeping a professional distance from the animals we studied. Although we were trained to develop animal models, we were warned against the hazards of explaining any of the behaviors we were investigating with a human explanation. This explanatory error is known as *anthropomorphism*, applying human characteristics to nonhuman animals. Darwin's student George Romanes, a naturalist, is known for committing the dreaded anthropomorphism bias by including tales of clever animals in his book *Animal Intelligence*. When your dog slumps off the couch when you enter the room and you declare that he is feeling guilty (the emotion you would feel after being caught in the act of doing something wrong), you're also committing this common mistake. Behaviorists can explain the behavior by emphasizing associations between the response of being on the couch and the predictable consequence of being punished, a much more parsimonious explanation than attributing the emotion of guilt to your canine friend.

For years, I've followed the marching orders of my graduate school mentors. As I have observed behaviors that have been conserved through mammalian evolution, I have qualified my suggestions of species similarities with apologetic statements in my lectures such as "Not to anthropomorphize, but . . ." or "I don't mean to anthropomorphize, but . . ." I'm beginning to feel that it's time to break away from my graduate school mandates and make a bold statement that, in some circumstances, the rats I examine are experiencing an emotion, motivation, or illness that is similar to that of humans. In fact, according to Elliott Sober, a leading scholar in philosophy of biology (at the University of Wisconsin at Madison), to deny such commonalities in cognitive capacities would constitute another, equally egregious philosophical error, one he calls *anthropodenial*.

As we examine common factors between humans and rodent mental capacities, it is important to avoid both overestimating (anthropomorphism) and underestimating (anthropodenial) commonalities. Consequently, we will let science guide us. Species similarities will be proposed based on a convergence of theoretical and empirical evidence. Before we consider advanced cognitive capacities, however, we have to create the neural circuits to support such endeavors.

Build-a-Brain: Deluxe Edition

I still remember the day I realized that the brain was the most exciting topic I had ever encountered. I was nineteen years old and taking what was known then as a physiological psychology course—the course with the reputation of being the most difficult in the psychology major. I teach that same course today, but it goes by the more contemporary name of behavioral neuroscience. Because I had a dual major in biology, the combination of biology and psychology promised by the course description sounded intriguing. But I had no idea that the information I would encounter in this course would shape my interests, passions, and perceptions for the rest of my life.

What was all the excitement about? My professor Janice Teal slowly revealed the intricacies of the brain in her distinctly southern tone and formal teaching style—as if she were sharing national security secrets. I listened as if my life depended on it. Up to that point in my experience, nothing had ever sounded as complex and important as this information. We learned about the historical attempts to discover more about the elusive brain tissue, covered the intricate details of the delicate building blocks of the brain known as neurons, and discussed the remarkable way these cells became activated so they could communicate with other cells in the nervous system. With nearly a hundred billion cells to corral into functional circuits in the human brain, it amazed me that any of us had the mental capacity to survive for even a day. But somehow the requisite connections were formed, the appropriate ions were shuffled across membranes, and just the right neurochemicals were secreted from tiny little buttons dangling from the neuronal extensions to generate a thought, an emotion, or a complex behavioral response. Simply amazing stuff.

But the pièce de résistance was the story about the rat brains. With her wisps of blond hair bouncing across her forehead as she shook her head for emphasis, Teal told a story about how placing young rats in a cage with lots of toys, a Disneyland of sorts, for just thirty days changed their brains, mental capacities, and behavior in unimaginable ways. This was better than magic—the heck with pulling a rabbit out of a hat, I had just learned that you could put rats in cages with other animals and stimulating toys and pull out a brain with richer connections and an enhanced capacity to learn. This work, known as the classic *enriched environment* studies, was conducted by a team of impressive young researchers at Berkeley. After hearing this lecture, the other courses of study in my liberal arts undergraduate education—British literature, differential calculus, principles of sociology—took a backseat to this exciting new discipline investigating how the context of our lives influenced the most essential component of our existence, our brains. My enthusiastic pursuit of this topic hasn't waivered for the quarter of a century that has passed since taking my first notes about the dynamic nature of the mammalian brain.

One of the innovative researchers in the early enriched environment work was Marian Diamond. I recently had an opportunity to speak with her about the research climate of that time, a time when researchers considered the brain to be immutable and dictated mostly by the directions of the genes. As we looked to our parents and relatives, the prospect that our neural fate was determined by our familial genes was more comforting for some than for others. But this new research suggested that our neural tissue is not carved in stone; it is molded by our experiences. Certainly, our genetic makeup influences important aspects of our neural development, but other factors also play critical roles in this process. As exciting as these results were at the time, the scientific dogma emphasizing the

> Our neural tissue is not carved in stone; it is molded by our experiences.

fixed nature of our brains was slow to change with the contemporary data. Diamond conveyed how scary it was being a young woman presenting such revolutionary data at an anatomy conference held in Washington, DC. She knew her results countered the current way of thinking about brains, but she had data to support her claims. Would that be enough? Apparently not. A scientist jumped up at the end of the talk shouting, "Young lady, that brain cannot change." Apparently, even for empirically minded scientists, new ideas are difficult to process.

The idea that the brain was fixed in form had not always represented the opinions of brain researchers. In the early nineteenth century, Johann Spurzheim, a sidekick of Franz Gall (the proposer of the interesting theory phrenology), was already thinking about the putative plasticity of the brain's tissue. According to the teachings of phrenology, individual differences in mental abilities were due to differences in certain areas of the brain (so far, so good), and the size of these brain areas could be assessed by feeling the size of bumps on the skull. Considering that most of us have fairly smooth skulls, the scientific merit gets lost with the bumps part of this theory. Despite the questionable existence of the proposed bumps on the skull, Gall and Spurzheim popularized this movement, which was quite big in the United States. In the nineteenth century, a potential employee may have had to sit for a phrenology exam before being hired. Spurzheim eventually split from Gall and started speaking of rather radical ideas for this time. He espoused ideas that mental disease could be treated by exercising compromised functions. Thus the radical idea of what would later be referred to as *neuroplasticity*—the notion that the brain continues to change throughout one's life and can be altered by our life experiences—had been introduced but it would be a while before there would be data to support these claims. As mentioned above, when the data did arrive on the scene, they weren't from humans, they were from rodents.

Following Spruzheim's innovative ideas, Darwin, in 1874, suggested that a complex environment was beneficial to the brain by stating that the brains of domestic rabbits were smaller than wild ones. Was there something about the challenges associated with living in the wild that stimulated brain growth? About a decade later, naturalist and children's writer Beatrix Potter made a similar observation when she wrote, "I had many mouse friends in my youth. I was always catching and taming mice—the common wild ones are far more intelligent and amusing than the fancy variety."

These early suggestions of environmental influences on the brain's landscape were still a long way from the laboratory. It is interesting that observations of pet rats played a big role in the study of enriched brains in the laboratory. Canadian psychologist Donald Hebb made a claim in 1949 that rats he had brought home as house pets performed better in learning tasks than the confined, caged rats in his laboratory. He speculated that there was something about the enriched living conditions in his home that led to better performance in the mazes. This sparked several behavioral studies confirming that, indeed, the enriched animals were better maze learners. At this point, the Berkeley team—including Mark Rosenzweig, David Krech, Marian Diamond, and Edward Bennett—stepped in to conduct the famous research confirming that enriched environments produced significant brain changes.

Decades of subsequent research have confirmed these initial

enriched environment results produced by the Berkeley scientists. The message is now clear that our brains do indeed change, and the environment we place our brains in plays a significant role in this process. One of the most celebrated effects includes increased connection points, known as spines, on the neuronal dendrites, the neural processes that extend from one neuron to another. But the impressive effects don't stop with increased dendritic spines; the list goes on: a thicker covering of the brain known as the cortex, larger cell body size of certain populations of neurons, more connection points between neurons known as synapses, more supportive cells known as glia, and increased capillary diameter, to name just a few.

Even more exciting is the observation that there don't seem to be any time restrictions for these brain modifications; at least there are no age restrictions. University of Illinois neuroscientist William Greenough has conducted numerous studies confirming this effect in adult rats. Further, if toys that provide opportunities for exercise are included in the enriched environment (as will be discussed in Chapter 10), additional blood vessels are observed in the brain. Thus, being exposed to novel objects, challenging tasks, and exercise appears to be a winning combination for the rat brains studied in these experiments, producing more complex neurons and an enhanced blood supply—essential ingredients for a healthy brain, regardless if you're a rat or a human.

The malleability of the brain, or neuroplasticity, is now accepted by all neuroscientists. Recently, enriched environments have been found to enhance the production of new neurons, technically known as neurogenesis. Certain therapies that eventually reduce the symptoms of depression, such as antidepressants and electroconvulsive shock therapy, also increase rates of neurogenesis. Could exposure to an enriched environment protect against the onset of depression or help reduce the symptoms of depression? Neuroscientists in India have shown that enriched environments protect against the toxic effects of chronic stress in rats. This is significant because it suggests that environments may rival, or even surpass, drugs in their ability to maintain healthy brains.

The Daily Grind

Most of us love our coffee breaks. Just give me a grande nonfat latte and a few minutes to regroup and I'm good for another several hours of work. In the midst of a busy afternoon at the office, there appears to be some benefit to taking a break away from the desk. This break doesn't have to include coffee—a walk outside, a brief conversation with a colleague, or just leaning back in your chair oriented away from the computer screen. On a more extreme level, some cultures take a longer break, a siesta, to escape from the hassles of the workplace.

Even rats placed on certain work schedules are known to take a type of break. In the laboratory, this break is known as a *postreinforcement pause*. The famous Harvard behaviorist B. F. Skinner provided evidence half a century ago that rats placed in chambers where they had to press a bar a certain number of times before receiving a reward (that is, a fixed ratio schedule similar to some assembly-line work) paused after receiving the reward and before initiating the next set of responses. The harder the rats had to work for their reward, the longer the pause. Humans also press bars in predictable fashions to obtain rewards—casinos in Las Vegas are filled with examples—but that's another story.

A recent study conducted by David Foster and Matthew Wilson at the Massachusetts Institute of Technology suggests that giving someone a break may be more than just restful, it may ramp up learning and neural processing. In this study, the focus was on the hippocampus, a brain area involved in learning and memory. If you've had a general psychology course, you'll remember a case study famously known as H.M. who had part of his hippocampus removed in an attempt to mitigate his severe epileptic seizures. This treatment successfully treated H.M.'s epilepsy but at the extreme cost of losing his ability to form any new long-term memories. This case study prompted a storm of research in rodents, investigating the specific role of this structure in memory formation.

I spend a good bit of my research time focusing on the hippocampus. And I have to admit that there's no other brain area where I'd rather spend my time. Sometimes when I'm looking through the microscope at the circular patterns and clearly defined rows of distinctly different cells, I get lost in the moment. Several years ago I had an opportunity to speak with the famous neuroanatomist and researcher Paul MacLean who spent most of his time conducting research at the National Institutes of Health. MacLean is known for naming the area of the brain involved in emotional functions—he

Looking through the microscope is like entering a cathedral.

called it the *limbic system*. In our discussion about his early brain excavations, he told me that looking through the microscope is like entering a cathedral. I know what he means.

Because the hippocampus has been shown to be critical for learning tasks that involve sequential decision making, typical of tasks such as navigation, Foster and Wilson wondered just what those neurons in the hippocampus did during these pauses or breaks after a sequence of behavioral responses. They created a rodent track so that the rats would run laps and eat a reward after each lap. After consuming its food reward, the animal would wait for a variable time period before beginning the next lap. Behaviorally, the researchers noted that the animals would groom, whisk their whiskers around, or just sit still—much like we may look around, fiddle with our hair, or just sit quietly during our breaks. But what was going on in the hippocampus?

This is not an easy question to answer. It involves a good dose of patience and methodological sophistication. The researchers implanted very tiny recording devices into individual cells in this brain area. Just to give you an idea of how precise this technique is, the entire rat brain is the size of a small pecan, and the hippocampus is about the thickness of the pecan shell and the length of a fingernail across the two hemispheres of the brain. The individual neurons definitely require a microscope to view—their tiny cell bodies are approximately fifty microns in diameter, the width of an average human hair. This is precise surgery!

Once the electrodes are in place, subsequent electrical activity is recorded as the animal traverses various areas of the track so that *place cells* can be identified. As the name implies, place cells are cells that fire when the rats are in a particular place. Once the code for the sequence of firing patterns for the neurons was established, the firing pattern during the break could be interpreted. These rest-time firing patterns represented the same pattern of responses seen during the previous run, but in reverse order—especially when the rat just completed an unfamiliar run. It was as if the rat's brain were rehearsing what it had just learned during the brief break. This form of sequential replay of neurons has also been demonstrated during sleep sessions after learning tasks. When interviewed about this study, Foster speculated that the repeated rewind processing of an event provides an opportunity for other brain areas to interpret and store the information. If a certain terrain has predators, a continued activation of these place cells—with the appropriate emphasis in the brain's fear and vigilance centers—would be valuable for storing and consolidating long-term memories of the event. If the rat enters this area in the future, it will be appropriately cautious about any predators that may be lurking around. Thus more learning may go on during these rat breaks than during the tasks themselves. If this were the case for humans, it would help us understand findings in cognition suggesting that humans learn more efficiently during intermittent study periods than when engaging in one long extended study period. If you have three hours to devote to studying, these findings suggest that you are better off working in three one-hour study sessions (distributed studying) instead of sitting down for a three-hour marathon study session (mass studying).

Before moving on to yet another impressive demonstration of cognitive processing in rats, I have to address the importance of the coffee part of the coffee break. Do we experience an added cognitive benefit when we accompany our breaks with a cup of coffee? It looks like the jury is still out on that question. In general, caffeine, a cognitive stimulant, has been associated with some neuroprotective effects in Parkinson's disease models, and it can certainly increase alertness and sustained attention that may enhance learning in the short term. A recent study, however, suggests that moderate doses can suppress neurogenesis, the generation of new neurons. If this neurogenesis suppression effect generalizes to humans, other enriching activities in our lives may interact with caffeine in a way that may increase the production of valuable neurons or buffer against the caffeine-induced decrease observed in this study. While researchers sort out the exact long-term effects of caffeine, the most important lesson we've gleaned from these studies is that, regardless of whether they are accompanied by our favorite coffee drink, we need to include periodic breaks in our busy schedules—that is, if we want to remember what we're learning.

More Similar Than Any of Us Thought

Philosophy is an interesting intellectual endeavor. An entire discipline devoted to the act of thinking about, well . . . thinking. If ever there was a cerebral course of study, philosophy is it. Most of us would never think that our rodent friends could inform us about the epistemological activities espoused by the likes of Descartes, arguably the most celebrated philosopher and intellectual of all time. Could rats inform us about

How do you measure metacognition in a rat?

the nature of knowledge and the extent to which we are aware of the knowledge we possess?

Until reading about the creative work conducted by Jonathon Crystal and Allison Foote at the University of Georgia, I would have not even asked the question. I wouldn't have guessed that rats could contribute valuable information to this question, to the topic of metacognition—thinking or reasoning about one's thinking. How do you measure metacognition in a rat?

First, the rats were trained to discriminate the duration of two tones so that they could determine if a test tone was longer or briefer than the standard tone. They were taught to declare their choices by poking their nose in a designated compartment. If the rats correctly identified the tone as brief or long, a large reward was presented. An incorrect response, on the contrary, produced no reward . . . not even partial credit. Rats had previously been shown to master discriminative learning tasks but what was different about this study was that, on some trials, another option was presented. The rats were given an opportunity to back out of the forced choice situation where they were required to make a correct response to receive the large reward. If they opted out of these forced choice trials, however, there was a consolation prize—the rats received a small reward every time they placed themselves out of the forced choice trial. Do you see what the researchers are doing here? If the rat is aware that he knows the answer, the smartest response is to play the game and take home the large reward. But if the rat has any doubts about the accuracy of the response, the smartest move is to back away and take the smaller prize.

The researchers made two predictions that would support the notion that they were demonstrating metacognition. First, the rats should take the smaller reward in the tasks that were more difficult to discriminate. Second, in those more difficult trials, the rats would perform more poorly when they weren't given the opportunity to opt out of making a choice. Both rat response strategies supported both predictions . . . they skillfully used the option to decline making a response when the test question was difficult but consistently played the game when they appeared to know the answer.

You may be thinking that we're dancing around those *anthropo-* terms I introduced earlier in this chapter. The authors boldly published this study with the title "Metacognition in the Rat," claiming that this was the first evidence supporting that a nonprimate possessed knowledge of its own mental or cognitive state. Regardless of what we call it, this study confirms that rats are processing information about the environment (considering the repetitive neural sequencing described in the preceding section) and their own thought processes in extremely sophisticated ways.

The Crystal and Foote experiment epitomizes how scientists can design studies involving rats to ask very interesting questions about seemingly human qualities. Their challenge represents one of the most rewarding aspects of working in the area of neurocognition: determining how to use a particular task to ask rats a rather sophisticated question. I can't help but be proud of the fact that these philosophizing rats are from the University of Georgia, where I had my very first encounter with real rats as a graduate student in biopsychology on the very same floor where these honor roll rats demonstrated their impressive intellectual capacities.

As I consider the implications of the rat metacognition study, it prompts a little *self*-metacognition as I think about past experiences with my human students. One of the challenges of being a professor is to provide the tools and motivation that allow students to be able to master the information and thus enter the classroom on test day confident they know the material. It's very discouraging when there's a mismatch between a student's confidence that she knows the material and her ability to demonstrate that knowledge on the exam. Students often comment on how they knew the information, which prompts me to scratch my head as I hand them a low test grade. In an attempt to delve a little deeper into the metacognition abilities of my undergraduate students, I've implemented a tactic for several years in which I give students what I call "coupon points"—maybe five points that they can use anywhere on an exam. This helps relieve some of their stress because they have a bit more control over the situation by, like the rats, being able to opt out of a few questions they feel they can't answer (they don't realize that I've made the exam a bit more difficult to compensate for this option). This testing situation gives me an opportunity to see just what my students know about what they know. Are they able to use the points appropriately? Some students lack a clear idea of the answers they do not know, missing several questions but opting to bank their points for the next exam, even though they know nothing about how difficult that test will be. Others correctly answer the questions, but use their free points anyway, meaning that they don't have a clear sense of what they *do* know. And it is interesting that others become so distracted with the whole process of assigning free points that they end up doing worse than if they hadn't received any free points. Students who have successfully studied for the exam are confident about what they do and don't know and use more effective strategies when making decisions about using their test coupons. There is a lot for all of us to learn about assessing the accuracy of the knowledge we possess; perhaps the rats will reveal additional insightful clues about these mental processes.

As we leave this chapter, I want to revisit the enriched environment rats. In her book *Enriching Heredity*, Diamond reported that a teenager in Shanghai once asked her about brain differences between creativity, typically observed before the age of forty, and wisdom, typical of a person's golden years. If the brain were more responsive to change during youth, she asked, how could older people be wiser? Diamond speculated from her rat studies that the pattern of dendritic spine enhancement may be the key component to gained wisdom. When she placed young rats in enriched environments, she found increases in dendrites closer to the cell body, which may be associated with functional modifications such as more focused creative activity. Older rats, on the other hand, exhibit the most dramatic changes in the most distal dendrites, allowing the neurons to communicate with a more diffuse array of surrounding neurons. Having the ability to activate more expansive neural networks could contribute to the ability to make more informed, wiser judgments.

Diamond's curiosity about the neuroanatomical underpinnings of intelligence and wisdom prompted her to venture out to the human brain. She was interested in the number of glial cells—the cells that deliver nutrients to the neuronal cells—in the human brain's advanced prefrontal cortex and other association areas in the cortex. Her rat studies indicated that higher numbers of glial cells to support the neural cells in certain brain areas were characteristic of the most highly evolved brain areas. After observing brain tissue from men ranging from their forties to their eighties, Diamond made a bold request to receive samples of Albert Einstein's brain. Pathologist Thomas Harvey obtained Einstein's brain after his autopsy in 1955. A quarter of a century later, a few square samples of several cortical areas were sent to Diamond's laboratory. Building on her rat brain work, Diamond identified at least one key difference between Einstein's brain and those of the other men. Einstein had more glial cells per neuron, especially in the region known as the *inferior parietal cortex*, an area that receives the most expansive array of information from across the brain, kind of an uber-association area. Alas, empirical evidence (speculative as it is since it is based on a case study) reveals similar underlying mechanisms of enhanced cognitive abilities in both rodents (Diamond's rats) and Nobel Prize–winning physicists (Einstein).

Darwin's proposed evolutionary continuum encompassing all mammals is confirmed with insightful and informative studies such as the research described in this chapter. As scientists absorb these findings, they increasingly view the use of appropriate animal models as an important and essential step toward understanding the central truths surrounding our own prized neural networks.

CHAPTER 3

|||||||||||||||||||||||||||||||||||||||||||||||||||||||||||||||||||||||||||||

Neuroeconomics and
Long-Term Investments

I often show my students a short video clip of a mother rat that I have in my coveted collection of animal behavior videos. In this case, the rat is a resident of an inner-city sewer system. Her nest, where she spends a considerable amount of time nursing and protecting her pups, was built in what appears to be a safe area away from the water. But when the water level suddenly starts to rise, the maternal rat is presented with the challenge of protecting both herself and her offspring. The mom gently cradles one pup with her long front teeth, carrying it to safety. With the water continuing to rise, you don't expect to see her again. But there she is—as long as she can navigate the rising water, jumping and balancing with the agility of a superhero, she continues to rescue her pups, one by one, risking her life with each additional trip.

The video clip is fascinating. This rat makes it to safety after her first escape attempt and even has a pup to take care of out-side of the hazardous sewer system. But as rewarding as safety and a healthy pup may be, certain activational patterns in her brain motivate her to leave both her pup and safety to brave the dangerous waters to retrieve another of her offspring. Pup retrieval is not an automatic response for a female rat. For exam-ple, virgin rats typically exhibit behavior that suggests that they find pups unappealing, maybe even repulsive.

Mating, pregnancy, birth, and lactation drastically change the reaction to pups that were once of little significance to the female rat. Certainly the cost of leaving the pups behind is greater for a maternal rat than for a virgin—those pups repre-sent her genetic investment. But how does an increased genetic investment get translated into such a dramatically altered behavioral response? Do the sensory stimuli—the sound, smell, and touch of the pups—now affect the brain's decision-making areas in qualitatively different ways? Is the information treated differently once it enters the brain? Have new family-value neu-ral circuits been constructed? Even though we're dealing with rats and rat pups, this is far from a trivial question. Unraveling these mysteries can inform humans on many levels. In this chapter we'll explore how the brain uses past and current infor-mation to make decisions when various options are presented—in more technical terms, this emerging discipline is known as *neuroeconomics*.

Essentially, neuroeconomics investigates the brain's role in decision making. When someone offers you a piece of cake, that's typically easy—you take the cake. But when you approach the dessert bar at your favorite restaurant it becomes more complicated. How do you decide which dessert to select?

Although the name of this new discipline suggests that the content of the field focuses on decisions about financial investments, it is much more than that. Each of us face a plethora of choices every day . . . it is interesting to consider the rationale to our responses. Why did you pick that particular type of toothpaste? Why did you agree to marry him? Why didn't you hire that applicant with the impressive résumé last week? There is no denying that your brain held your mental hand in these circumstances, but just how did it do it?

Neuroscientist Paul Glimcher at New York University has conducted groundbreaking work in this area with primates. His work suggests that our responses, especially in indeterminate situations in which there is more than one response option, are more complex than the simple sensory and motor input and output described by Descartes and other pioneers in brain science. The key brain area implicated in decision making, the cortex, receives multiple messages from the environment, the body, and other brain areas. Thus for each and every decision you make throughout the day, your brain interprets information about the current environmental context in tandem with relevant physiological processes, to direct your final response choice . . . Coke or Pepsi? Papa John's or Domino's? Quality time with your child or a bubble bath? Using elegant studies with primates, Glimcher has identified certain areas in the most recently evolved cortex that appear to play a role in effortlessly calculating the costs and benefits of various responses. Although you may not have performed well on that probability exam in your statistics course, your brain has been successfully using probability and game theory for your entire life.

Will Work for Rat Pups

Before the formal introduction of neuroeconomics, however, behavioral neuroscientists were presenting rats with interesting choices in an attempt to learn more about the value of one stimulus over another. Alison Fleming, working at the University of Toronto at Mississauga, has conducted several ingenious studies to assess the value of pups to mother rats. Were they worth working for? She answered this question by training female rats in what is known as an *operant chamber*, first designed by the well-known behavioral psychologist B. F. Skinner (introduced in Chapter 2). In this experimental situation, female rats were trained on a simple schedule in which they were rewarded with a piece of Froot Loops cereal every time they pressed a bar located on one of the walls of the chamber. If you can imagine being very thirsty as you approach a vending machine that contains just one type of soda in it and receiving a drink every time you press a bar, you've got the idea of a Skinnerian operant chamber. The *work schedule* can get more interesting by requiring the rat to press the bar several times to get the reward, but for the purposes of this study, each press of the bar was rewarded. The rats' persistence on the task revealed their motivation to receive Froot Loops rewards. Then, after the female rats had given birth, Fleming and her colleagues switched the task up a bit. Now when the rats pressed the bar a new rat pup gently slid down the shoot. I won't ask you to imagine such a vending machine, but you get the point. As was the case for Froot Loops, the maternal rats kept pressing the bar as the chamber slowly filled with helpless rat pups. These maternal rats seemed to genuinely value young rat pups, evidenced by their willingness to work for the payment of a pup.

Scientists have also used a technique called *conditioned place preference* to assess the value of certain stimuli by introducing the choice option important for the discipline of neuroeconomics. In this situation, rats are exposed to different chambers that are associated with different conditions. One chamber may have a chocolate scent and the other the scent of another rat, perhaps of an attractive rat of the opposite sex. In this type of study, scientists simply record how much time the rats spend in each chamber. The chamber in which the rats choose to spend the most time is deemed the preferred chamber. Using this task, Fleming simply asked the rats what they found more preferable: a chamber with rat pups or an empty chamber? The rat moms chose the rat pups, confirming the value of these young animals to the maternal rats.

But was there a limit to this seemingly boundless adoration for helpless, needy rat pups? Joan Morrell and her colleagues at Rutgers University found that, indeed, there were such limits. Morrell used the conditioned place preference test to determine the rat's choice between rat pups and a hit of cocaine. (The rats had experience with both conditions during training for the test—they were exposed to the pups in one chamber and, in a distinctly different chamber, experienced the psychoactive effects of cocaine.) This is getting interesting, isn't it? Without the social pressures to do the right thing that would be seen in human studies—not to mention the ethical considerations of giving young moms hits of cocaine in special rooms—this clever experimental design provides an opportunity to look at raw mammalian predispositions.

Morrell's data indicated that the maternal rats chose the pup chamber over the cocaine chamber when the pups were young, in this case up to about eight days of age. The tables turn, however, when asked to choose between cocaine and older pups of about sixteen days of age. In this testing situation, the moms go for the cocaine. Thus younger offspring are more attractive to rat moms, and perhaps other mammalian parents. That attractiveness wanes, however, as offspring become more independent. The older pups weren't quite rat adolescent age, but mothers of teens often express a maternal kinship of sorts with the rats when I tell them about this study, as if they could imagine that a hit of cocaine would sometimes be more appealing than dealing with the challenges of their teenage children!

These rat mom tales provide clues about the changing value of various stimuli in our worlds. It is interesting that the incorporation of work as a tool to reveal information about the perceived value of various resources in these types of studies has led to new research questions. Now that we're learning more about the brain's intricate connections between work and rewards, for example, scientists can assess the effect of removing the opportunity to work from an animal's behavioral regime. Not having to work is associated with living in the lap of luxury in many human societies, but are there costs associated with the coveted lifestyles of the rich and famous? The rats apparently have some answers to share about that very question.

> Are there costs associated with the coveted lifestyles of the rich and famous?

The Trust Fund Rats

I remember standing in my kitchen one Saturday afternoon having a phone conversation with my younger brother who was struggling with a deep personal issue after the death of our mother. Adding to his grief, he was facing the challenges of coming to terms with his sexual orientation—a stressful experience for someone raised in a small, conservative southern town. Although he hadn't been officially diagnosed, he was experiencing symptoms of depression—it was difficult to get through life's daily tasks; the joy with which he had typically approached many areas of his life was slowly diminishing

In an attempt to resolve this emotional conflict, my brother, who is a medical doctor, sought the help of a reputable psychiatric clinic. The clinic was very expensive but if their services helped resolve the emotional pain he was experiencing, it was thought to be worth it. After the two-week treatment, including isolation interspersed with talk sessions with psychiatrists, my brother called to give me an update. His sessions sounded like they were distinctly psychodynamic in nature—*psychodynamic* is the term used to describe theories or treatments derived from Sigmund Freud. Without going into all the details, let me just say the empirical evidence is lacking in Freud's attempts to identify salient factors leading to emotional illness.

My brother left the clinic with memories of odd encounters with psychiatrists, a bill for thousands of dollars for services rendered, and an evaluative letter summarizing the problem. I still have a portion of the letter:

The patient underwent psychological testing. . . . It suggested an unconscious quest for sort of twin-like connectedness for his loss of real self which he relinquished in order to be the perfect son for his mother. To be perfect was to live up to the mother's idealization which he may well have lost once his mother was no longer available to provide reinforcement. Consequently, sexual yearnings and behavior may represent a search for a mirroring self to reinstate the equilibrium and validating responses. The patient was introduced to these dynamics but their complexity will require continuing work before the patient will be able to more fully utilize the importance of this information.

As my brother read this part of the letter to me, I felt my blood pressure rising. Perhaps I missed something, but what in the world does this psychobabble mean? What about all of the insightful research that has been conducted to learn about emotional responses, the many lectures conveying the importance of empirical evidence that are drilled into psychology students across the country, and the encouraging data suggesting that cognitive-behavioral therapies are beneficial modes of treatment for several mental health challenges? None of this appeared to be relevant here. This diagnostic summary was based on clinical insight, wisdom, and experience as opposed to empirical evidence necessary for the approval of more traditional medical treatment strategies. The evaluation letter may have sounded intellectual but the words had no relevance and meaning for a patient who continued to suffer. Sure, my brother missed his frequent interactions with our mother (so did I!)—no news bulletin there—but he needed solid, practical advice that would help him feel better, not psychoanalytical prose. Before thinking about it, I blurted out that I felt that he would have been better off digging potatoes for two weeks.

You might guess that I had no empirical data to support my potato-digging therapeutic claim, so, in that sense, I was in no better position than the psychiatrist I had just criticized. However, with the help of some rats and the guidance of a rich scientific literature suggesting that physical effort used to obtain meaningful rewards is valuable in keeping the brain active and healthy, I went to the drawing board to design my potato-digging study.

The Rat Pack Starts Digging

As you learned earlier in the chapter, rats will work for Froot Loops. Instead of potatoes, I thought perhaps we could ask rats to dig for Froot Loops. My students built a square apparatus— about two feet by two feet—and covered the bottom with the bedding used in the rat cages. Rats were trained to dig in randomly placed piles of bedding that contained pieces of the sweet cereal rewards. This routine of harvesting Froot Loops continued for five weeks. Before we knew it, our dedicated rat workers would explore the entire apparatus for mounds, harvesting all the Froot Loops in a short period of time. I was interested in the value of these daily experiences in building strong associations between directed effort (work) and rewards in life (Froot Loops). We designed a challenge task by taking a ventilated plastic cat ball (a pet toy) and placing a large piece of Froot Loops in it so that the rat couldn't get it out. But the rat didn't know that. We

> We were interested in how long the rats would keep trying the task before giving up.

were interested in how long the rats would keep trying the task before giving up—making the decision that, regardless of effort directed toward the task, they couldn't get the reward is in many ways reminiscent of the behavior demonstrated by clinically depressed humans.

To test our hypothesis that this training—what we called *effort-driven reward training*—was beneficial, we needed a control group to be sure that all of the effects we were observing were not just a function of eating Froot Loops. So we created another group that would be placed in the digging apparatus for the same amount of time each day with the same number of mounds of bedding. The only difference was that this group didn't have to work for their Froot Loops; they were given their sweet rewards regardless of whether they expended any effort to obtain them. In scientific terms, the group that had to dig to retrieve their rewards was the contingent group (reward was contingent on effort), and the group that received the rewards regardless of effort was the noncontingent group. However, in the lab, they quickly became known as the worker versus the trust fund rats (as mentioned in Chapter 1).

> In the lab, they quickly became known as the worker versus the trust fund rats.

After weeks of daily training, we were ready for the big test. Would daily work efforts enhance emotional resilience in the form of persistence in these animals? Was there a brain boost to having a strong work ethic? We were thrilled to see that the worker rats spent about 60 percent more time working on the task than did their trust fund counterparts. We have replicated this several times, each time with similar results. In addition, the worker rats spend about 40 percent more time directed toward the task on each attempt to retrieve the cereal before walking away. Could it be that the worker rats, with more clearly defined associations between their efforts and obtained rewards in life, had a perception of being better able to success-fully complete the task—something psychologists typically refer to as *self-efficacy*?

The results of this intriguing study made me wonder about the impact of effort-driven reward training on the rats' brains. After reading some interesting research on the effects of neu-ropeptide Y (a chemical that naturally exists in our brains) on emotional resilience, I decided that would be our starting point. If this substance is infused in the amygdala (known as the fear center of the brain), animals exhibit less anxiety dur-ing a social interaction test than do control animals who do not have the infusion. Further, in humans, plasma levels of neuro-peptide Y were found to be higher in special forces soldiers exposed to a stressful prisoner-of-war experience than in non–special forces soldiers. If effort-driven reward therapy enhances resilience in the rats, could it also enhance resilience neuro-chemicals, without the use of psychoactive drugs such as anti-depressants? We trained another group of animals, exposed their brains to an antibody for neuropeptide Y, and found a stronger presence of neuropeptide Y in the brains of the worker animals. Encouraged by these preliminary results, we're currently looking at a resilience hormone known as DHEA (much easier to say than dehydroepiandrosterone) because worker rats have higher peripheral levels, as measured in their fecal in this case, of this hormone as well.

I admit that I haven't definitively resolved the question of the potato-digging therapy being more effective than the psychodynamic approach used on my brother, but evidence is building to support the value of effort-driven rewards. My brother bypassed further psychiatric assistance and opted to incorporate his favorite activity—cooking for friends—into his life routine. He finds this gourmet therapy to be extremely important for his mental health, and his friends who are often dinner guests fully endorse his nontraditional form of therapy.

Our rat cousins are demonstrating an important life lesson: Work may not only be beneficial from a financial perspective but may also lead to healthier brains that are resilient to the many challenges life throws our way. We'll discuss other forms of rat therapy in Chapter 5.

Healthy Brain, Healthy Work Ethic

Pioneering neuroscientists James Olds and Peter Milner, working at McGill University in the 1950s, identified the brain's equivalent of pay dirt when it comes to reward systems. Rats with tiny electrodes implanted into certain areas of their brains would work to complete physical exhaustion to press a bar delivering a mild electrical current to electrodes implanted in their tiny brains. In the 1960s, researchers at the Walter Reed Army Institute of Research reported that one rat pressed the bar to receive brain stimulation almost continuously for twenty days, averaging 42,106 responses per day.

That's certainly a strong work ethic! Although the brain area in this study was in a more traditional motivational area known as the *lateral hypothalamus*, most of the research directed toward the brain's pleasure, or reward, center has targeted a small forebrain area known as the *nucleus accumbens* and the neurotransmitter it uses, called dopamine.

Recent research involving more natural modifications in the brain's reward system, however, has revealed a much more complicated story. Instead of flooding the system with unnatural levels of dopamine caused by electrical activation, scientists currently insert tiny probes that assess the natural flow of dopamine in the targeted reward centers. Dopamine located in the reward center can be chemically altered with various drugs to determine the willingness of the rat to work when dopamine is or is not present. University of Connecticut neuroscientist John Salamone, as well as other neuroscientists around the world, have conducted many elegant studies to determine just what rats find valuable in the operant chamber tests—approaching the bar, pressing the bar, or eating the sweet reward. Applying these questions to ourselves, we may venture that we enjoy the paycheck more than the work itself, but when given thoughtful consideration, the answer may not be that simple—for us or for the rats. Generally, these studies tell us that dopamine-related signals increase during lever pressing and then drop during the consumption of the food reward. These results suggest that the brain reward system may be more closely involved with anticipation, exhibiting maximal activation while the animal is preparing for or working toward a goal rather than actually receiving the big payoff.

Consider the findings of another rat work ethic study. After depleting the rats' brains of dopamine, the animals were asked to navigate a type of obstacle course to receive a highly valued food reward. Whereas control rats who hadn't had their brain chemicals altered climb barriers to obtain a yummy food reward, animals with depleted dopamine circuits aren't as willing to work for the high-value food, especially when a low-value food requiring less effort is readily available. Thus if a rat is placed on a work schedule requiring minimal effort to obtain its rewards, reductions in brain dopamine or accumbens activity may not have an effect on its work patterns. When the schedule demands are higher, though, requiring the rats to work harder to earn their rewards, those whose brains have diminished activation capacity in the reward center were less motivated to put in extra work to gain more desirable treats. They seem to be perfectly happy picking the low-hanging fruit. Applied to my own profession, low mood states such as sadness or depression may not affect the daily activities of teaching my classes, but they would likely decrease my efforts directed toward designing exciting demonstrations for my classes or conducting original research projects, writing manuscripts, or applying for grants—all of which require lots of extra work but could result in a handsome reward of enhanced student interest, exciting new data, or the funds to conduct further research.

The neurochemical dopamine is prevalent throughout the nervous system and is strategically involved in the executive center of the brain, known as the *forebrain area of the cortex*. The forebrain has been implicated in effort-related decision making; if dopamine in the brain's pleasure center is disrupted, the cost–benefit analyses related to work and rewards are also disrupted, as just described. This putative reward system may be better characterized as a brain-activation system, propelling us to complete the required level of work to obtain relevant rewards in life. Perhaps Stanford economist Tibor Scitovsky was aware of this information, revealed by our hardworking rats, when he first wrote *The Joyless Economy* in the 1970s. He described the hazards of focusing on a lifestyle emphasizing comfort over rewards:

> We saw that while comfort hinges on the level of arousal at or close to its optimum, pleasure accompanies changes in the level of arousal toward the optimum. That is why the satisfaction of a need gives both pleasure and comfort. But the continuous maintenance of comfort would eliminate pleasure, because, with arousal continuously at its optimum level, there can be no change in arousal toward the optimum. In other words, incomplete and intermittent comfort is accompanied by pleasure, while complete and continuous comfort is incompatible with pleasure.

Considering how important work and activation are to the brain, it makes us question the wisdom accompanying the notion that we experience a more pleasurable lifestyle by hiring others to do our work—cut our grass, cook our food, clean our houses, and so on. Maintaining our bodies and brains in the comfort

THE LAB RAT CHRONICLES

Maintaining our bodies and brains in the comfort zone may be the worst strategy for increasing pleasure.

zone may be the worst strategy for increasing pleasure—that is, according to the rats. Taking sips from our water bottle at an optimal frequency to ensure that we're never thirsty deprives us of that intense satisfaction we feel when we take that first sip after a long stretch of having nothing to drink. And considering the results of the trust fund study described earlier, our cushy lives may also affect our emotional health as well. When Justin Joffe designed a neighborhood for rats that required them to work for their resources (such as water, food, and light), he found that the working-class rats exhibited less anxiety than the rodents who were given all their resources free of charge.

So, rat ventures into the brain's pleasure centers have produced some rather informative results. Neuroscientist Kent Berridge, mentioned in Chapter 1, has emphasized the brain's accumbens–dopamine circuit during states of wanting, which is a different psychological experience than the more hedonic sensation of liking. His clever rodent research suggests that a healthy wanting system that clearly understands the relationship between effort and rewards is essential to successful, adaptive living. This life lesson, learned early and practiced consistently, could well be the difference between a life lived versus one lived fully.

Diversifying Our Financial Portfolios

Although rats don't invest in the stock market or consider careers as venture capitalists, they can inform our work behavior in many ways. In fact, the studies described in this chapter clear up a lot of the confusion in our human world of work. To benefit from this valuable research, we need to reconsider the true value of the traditional rewards of our work: money, comfort, and decreased physical effort. As Emory neuroscientist Dennis Choi suggests, humans didn't evolve just to sell derivatives. Have we taken a wrong turn in our attempt to enhance the pleasure in our lives?

Nowhere is Choi's suggestion about our brain's difficulty evaluating financial derivatives more appropriate than in the 2008 crash of the U.S. stock market. As told by financial journalist Michael Lewis in his book *The Big Short*, it appears that the average human forebrain was ill-prepared to process the true value, risks, and costs associated with real estate derivatives. As disconnections between effort (cost) and real value (reward) became more prevalent, consumers continued to go to the reward bar in their financial versions of a Skinner box— they kept buying into mortgage-backed securities that they didn't fully understand because the reward seemed to be too good to resist. The rats exhibited similar distorted response strategies when electrodes were implanted into their brains, but in the case of the mortgage crisis of the last few years, no neurological alterations were required for the maladaptive responses that led to the meltdown of our economy. Rodent studies suggest that clear connections between expended effort and subsequent rewards are necessary for the most adaptive behavior to occur. If established effort–reward contingencies become too abstract or distorted for the pleasure and executive functioning areas of the brain to function properly, it may be a good time to walk away.

These results explain the recent findings of another Emory neuroscientist, Greg Berns, and his research team. Working with humans, he collected functional MRI (fMRI) readings (brain scans indicating relevant brain activity during certain tasks) from subjects asked to make a series of financial choices. In one group, advice from a financial expert was provided. He found that the group receiving the expert advice showed less activity in their brains than the group of subjects responsible for making their own decisions. So, following the advice of experts causes parts of our brains that would normally be involved in the process to shut down (or at least become less involved than expected). In the context of the mortgage crisis, reputable agencies such as Moody's and Standard & Poor's, the industry experts, provided their highest ratings to fabulously risky mortgage-backed securities. This explains a lot.

> The brain's reward center appears to be more engaged as we approach the simpler pleasures in life.

Thus the true rewards in our lives are composed of more than stock portfolios and net worth values. The brain's reward center appears to be more engaged as we approach the simpler plea-sures in life . . . food, sex, social support, and good old-fashioned work that produces tangible rewards. As these connections become increasingly more abstract and the rewards in our lives become less tangible, the integrity of the neural networks mod-ulating our behavioral responses become compromised.

Because part of my research program is related to investi-gating animal models of depression and resilience, I often Google the word *depression* just to see what has been recently reported about the topic. In the past, I became frustrated when I had to weed through the other depression—that is, the finan-cial crisis of the 1930s, a topic I perceived to be far from my neuroscience interests—to get to my mental health sites. Erro-neously, I failed to recognize any meaningful connections between an economic depression and emotional depression. But as I consider what the rats tell us about the value of work and its relationship to our mental health, perhaps there are more connections than I ever imagined. The data tell us that one of the sole purposes of the brain is to keep us interacting with the world around us to provide a consistent supply of the necessary resources for survival. During the hardships of the Great Depression when unemployment rates reached 25 per-cent, the inability to work to achieve life's resources was no doubt a devastating blow to the brain's circuitry. In addition, losing a clear sense of association between effort and reward, all too evident during the Great Depression, compromised the brain's ability to confidently make strategic decisions about profitable business options and investments. Consequently, as a nation, it took quite some time to recalibrate our collective work-related circuits. According to the rodents, the recent Great Recession may yet be another wake-up call—reminding us of the underlying mechanisms of our brain's decision-making abilities when it comes to getting the things we most need and want.

||||||||||||||||||||||||||||||||||||||||||||||||||||||||||||||||||||||||||||||||

CHAPTER 4

Universal Healthcare

In the United States, the political talk shows have been dominated by the topic of universal healthcare for the past decade. We are obsessed with worries about our medical options once we're sick and want assurance that we are eligible to receive the best treatment the medical establishment has to offer. When we consider, however, that although we spend more than twice as much on healthcare than any other nation and the fact that the United States ranks twenty-fourth among other nations for mortality rates in men and thirty-first for women, it makes one wonder what all the hoopla surrounding this so-called healthcare is all about. Sure, the medical doctors' ability to conduct precise surgeries at the microscopic level, swap organs, and train viruses to interact with our genes is impressive, but for the major killers out there (cardiovascular disease, stroke, pulmonary disease, diabetes), such high-tech medicine has not suppressed the emergence of deadly medical challenges.

When I look to the rats, I often think about what it would be like to have no external health assurance or insurance—to be responsible minute by minute for your health. With no promise of help once an illness emerges, the emphasis has to be on prevention. Indeed, the rats' universal healthcare system involves physiological systems that are keenly sensitive to toxicity and danger, especially when it comes to determining what is safe to put into their bodies. In this chapter we'll discuss the wild rodents' heightened vigilance about food consumption as we try to understand where humans have taken a wrong turn toward uncensored consumption, leading to deadly obesity. On the other hand, when rats are given all the food they want and not provided an opportunity to exercise, their eating habits and health consequences morph into a pattern that is very familiar to all of us . . . another reason why the rat is an excellent model for humans. And while we're on the topic of censoring what we put into our bodies, we'll consider how a little-noticed study conducted in the early 1980s may have provided the most valuable piece of the drug addiction puzzle.

Even with the best threat detectors, both rat and human bodies are frequently penetrated by nasty pathogens, resulting in illness. Thanks to some revolutionary rodent research in the 1970s, we've learned that the immune system has a few tricks up its sleeve that lead to enhanced healing. There are indeed many ways, sans drugs, to tap into this complex immune system.

The Most Important Ingredient for Healthful Eating: An Ounce of Prevention

As described in Chapter 1, the similarities between the eating habits of rats and humans make these rodents a valuable animal model for humans. We're both omnivores, attracted to a wide variety of foods. But with each new gastric opportunity, a new chance of contamination presents itself. Described as the *omnivore's paradox* by University of Pennsylvania psychologist Paul Rozin, both humans and rats find themselves in a perpetual quandary when it comes to their coexisting attraction (neophilia) and hesitancy (bait shyness) for consuming novel food. Perhaps more than humans, rats err on the conservative side. If a new food substance is placed on a frequently traveled path, they will bypass the new food for quite some time. After a while, the rat will take a small portion and consume it. If no sickness follows the consumption, then the rat adapts its diet to include the new food substance, explaining why inner-city rats happily consume pizza, fried chicken, and Chinese food. Providing further evidence of their uber-vigilant threat-monitoring systems, if a familiar food is placed on a familiar path, but something else just isn't quite the same (for example, a rock is in a different place), the rat is hesitant to consume the food.

As I think about the vigilance these rats have about their food consumption, I'm trying to remember the last time I bypassed a yummy-looking dessert because I had never eaten it before. And I'm quite certain I've never passed up a snack because something looked out of order on my kitchen countertops. Other than taste incentives, there appear to be no neural processes dedicated to the task of monitoring my eating habits—a food-regulating neuronal wasteland! Apparently this is not a novel observation. Nineteenth-century Harvard experimental psychologist William James, known as the father of American psychology, reported:

> Not one man in a billion, when taking his dinner, ever thinks of utility. He eats because the food tastes good and makes him want more. If you ask him why he should want to eat more of what tastes like that, instead of revering you as a philosopher he will probably laugh at you for a fool.

If we dig deep enough, however, it is evident that humans share a few of the rodents' vigilant systems. A famous study in the 1950s shed some light on one of nature's insurance policies against consuming poison. In fact, I'd say nature devised a program of comprehensive coverage against being poisoned, a program that you will recognize as being used by humans.

John Garcia's discovery of this impressive health insurance policy was quite serendipitous. Tired of graduate school at the University of California at Berkeley, he accepted a job with the military before completing his doctorate. In his first real-world position, he was instructed to investigate the effects of irradiation on biological systems. He housed rats in cages and, after exposing them to irradiation, assessed various physiological systems—pretty run-of-the-mill, mundane research. But Garcia noticed something kind of odd about the rats' behavior after they were placed back in their home cages. When water was placed in plastic bottles instead of the usual glass bottles, the rats avoided drinking. Could it be possible that just a single trial of irradiation exposure (and subsequent illness) created a strong aversion of variables related to the sickness? Consuming water could easily be associated with an upset stomach, and the rats were quick to associate the plastic-tasting water with their illness; maybe there was something to this observation.

Garcia needed to test his hunch that the rats learned to avoid cues associated with illness, and he did so by determining if rats would associate saccharin water present during gamma radiation with their subsequent sickness. If it worked it would be fascinating; rats are attracted to sweet water—could this attraction be turned into an aversion? The test was easy. We know that the typical rat prefers the saccharin water to plain water, so if the rat avoided the sweet water in a two-bottle taste test, then it would provide strong evidence that a conditioned, or learned, aversion had developed between saccharin and illness. Of course, the saccharin hadn't actually caused the illness, but the rat didn't know that. It was certainly a logical conclusion to assume that the last item consumed led to the illness, and the last consumed substance was the water. In the rat's natural world, the combination of the bait-shyness tendency to consume only a small amount of a novel food and the strong associations formed between a specific food or taste and subsequent illness provided the animals with an invaluable survival strategy.

For the die-hard animal behaviorist, however, these findings were interesting for several reasons. Garcia's studies cleverly showed that the aversion could develop after just one pairing (that is, one-trial learning), and the time interval between the food exposure and the sickness symptoms could be extended to several hours. In addition, the associations were more readily formed between a taste and the gastric sickness than between a nonfood stimulus (such as bright lights or noises) and gastric illness. This finding made sense because a taste is most often a consequence of something the rats recently ate rather than something they recently saw or heard.

As interesting as these results were, the world of psychology wasn't ready to hear about these odd observations. The great physiologist-turned-behaviorist Ivan Pavlov had finally provided the discipline with some hard-and-fast rules of behavior relating to conditioning and learning. For learning to occur, several conditioning trials were necessary, the stimuli or events needed to be experienced very close in time, and there was no evidence that certain types of stimuli were better than others at eliciting learning. Garcia's accidental observations threatened these rules of the new discipline of psychology; consequently, it was difficult for him to publish his rat taste preference findings. Eventually, scientists realized the value of Garcia's discovery of conditioned taste aversion in rodents. And Garcia's findings addressed the frustration of professional pest-control workers—the conditioned taste aversion, coupled with their bait shyness, make it extremely difficult to coax a rat into eating a poisoned food substance.

The University of California at Berkeley eventually awarded Garcia a doctorate degree, and his discovery made him a valuable alumnus. When he visited my students at Randolph-Macon College almost two decades ago, his stories of sitting in the advanced learning classes at Berkeley were amusing. Garcia told the students that he really didn't fit into the scholarly student crowd when he was a graduate student. While everyone was espousing the learning rules du jour (in this case, the tedious postulates proposed by Clark Hull), Garcia was telling stories about trying to train his dog. It's interesting that his discoveries about species-relevant taste aversions remain a seminal finding in the field of psychology, while students read about Hull's postulates only in history books. This is a wonderful example of the elegance of research—sometimes the simplest, even accidental, observations can be the most revolutionary.

If you've ever experienced food poisoning, you've done your own research in conditioned taste aversions. Your avoidance of the specific food that made you sick, perhaps for the rest of your life, confirms the generalization of Garcia's rat findings to humans—more evidence that lessons we've learned from rats apply to humans. There is no doubt that this gastric security system continues to exist in humans.

I often think how beneficial it would be if food aversion extended to foods that would eventually lead to long-term health threats, such as obesity, diabetes, and hypertension. It would be wonderful if we could learn about the harmful effects of consuming a type of food or too much of a food and then make the decision to avoid such consumption preferences. Our lower brain areas that evolved to respond positively to tasty foods override our executive centers and maintain our attraction toward those foods. Even after being told that our life is threatened, we will continue to pull into the drive-through at our favorite fast-food restaurant. And this disconnect works both ways—after learning that unwanted bacteria, not the food itself, made us nauseous, it is virtually impossible to go back to liking that particular food again. The inability of our higher thought centers to regulate and strategize information related to our consumption habits is even more dramatic when it comes to consuming substances that are the most effective at activating the brain's reward systems—namely junk food and psychoactive drugs, the next topics of interest in our rodent's-eye view of universal health.

Junk Food: Even the Vigilant Rats Can't Resist

On a few occasions I have had the opportunity to meet Princeton University neuroscientist Bart Hoebel at neuroscience conferences. Hoebel stands out for two reasons: First, he's very likely the tallest neuroscientist I have ever met, and second, there's something about the way he conducts and conveys his research that suggests he has a unique ability to listen to his laboratory rats. His research questions are insightfully guided by his past observations; consequently, the rats have directed him to some of the most interesting and relevant research questions in the field of behavioral neuroscience, all dealing with addiction and food. Consequently, in many ways Hoebel is a scientist ahead of his time; that's why I never miss an opportunity to hear him speak, to soak up just a little of his impressive rodent wisdom.

High on Hoebel's list of innovations is the observation that rats become addicted to sugar. Because drugs of abuse activate the brain's pleasure center and neurochemistry in similar ways as food, the idea of food addiction is certainly plausible. Hoebel and his collaborators have demonstrated that rats will binge on sugar water and will exhibit signs of withdrawal (including anxiety, depression, craving, and even teeth chattering) during sugar abstinence. Perhaps the most disconcerting finding is that sugar-addicted rats consume more drugs of abuse during sugar abstinence than nonsugar-addicted rats. This finding suggests that seemingly harmless sugar may serve as a gateway substance for other drugs of abuse. This makes you think of those Twinkies in a whole new light, doesn't it?

> Seemingly harmless sugar may serve as a gateway substance for other drugs of abuse.

Further evidence of food addiction was recently revealed by researchers at the Scripps Research Institute in Jupiter, Florida. Researchers exposed rats to different versions of a cafeteria diet—a diet consisting of several tasty foods, typically shown to enhance motivation and weight gain in rats. If you've ever heard of the *freshmen fifteen*, a weight-gain phenomenon described by college students, this is similar to what is observed in the rats. Regardless of your species, having the opportunity to eat several types of high-incentive foods results in increased food consumption. Rats placed on the cafeteria diet in the Scripps study feasted on chocolate, cake frosting, pound cake, cheesecake, sausage, and bacon. Rats either had extended access to the junk food diet, limited access, or no access. For all groups, standard rat chow was also available. Confirming the notion of the freshmen fifteen phenomenon, the cafeteria rats consumed *twice* the calories consumed by the rodent chow group. Even in the group that was allowed to eat junk food just one hour per day, the rats consumed about 66 percent of their calories from the junk food. The rats truly developed a taste for unhealthful food; when the cafeteria diet was terminated, the rats refused to eat substantive amounts of healthful food for almost two weeks.

This study also provided evidence that the junk food rats' brains had changed in a pattern similar to that observed in heroin addicts. The receptors in the brain's pleasure center (the nucleus accumbens) that regulate the reward neurochemical dopamine were down-regulated (or underwhelmed), leading to a dependence on the drug—or food—for normal maintenance of the brain's reward system. Although acquiring a sweet tooth seems harmless, the research conducted on the rodents suggests that there are few distinctions between food and drug addiction.

Now that research has identified both rodent and human biological predispositions to consume certain types of foods, what impact does this have on advertising and marketing strategies employed to attract us toward often unhealthful food? It is interesting that as I was completing this chapter and pondering this question myself, a lawsuit was filed against McDonald's fast-food restaurants for including a toy in their Happy Meals. Based on what is known about food preferences and consumption habits, you should have a new respect for the cleverness of the McDonald's executives who initially thought of this ingenious idea. As children associate their delight prompted by the toys in the Happy Meals with the cheeseburger, fries, and soda, positive emotions will soon be formed so that when the child encounters McDonald's advertising or smells fries in the mall food court, a warm, happy feeling will emerge, leading them toward the Golden Arches. Add to that the fact that in many cases these associations can last a lifetime, and you've got quite a marketing plan! Of course, the losers in this situation are the children who are taught to associate McDonald's high-fat foods with positive emotions. It is not surprising that the lawsuit claims it is unethical to create these associations in children, especially considering the unhealthful nature of the typical Happy Meal.

Sarah Leibowitz's lab at Rockefeller University has identified additional parallels between food and drug consumption. Just as women are warned to avoid the consumption of drugs and alcohol during pregnancy, Leibowitz's research informs us that we need to add excessive consumption of high fats to those guidelines. On the positive side, the offspring of rats who consumed a high-fat diet during pregnancy exhibited increased rates of neurogenesis, the production of new neurons. More brain cells is good, right? In this case, the new neurons migrated to the area of the brain known as the hypothalamus and matured into cells that remained throughout the entire life of the rats. What did they do? They generated an appetite for a high-fat diet. Perhaps, in this case, those additional neurons aren't so good after all—and I'm wondering if I have an especially high number of those neurons in my hypothalamus!

Marian Diamond, the Berkeley neuroscientist who conducted the classic enriched environment studies described in Chapter 2, recently told me about some studies she conducted suggesting that an impoverished eating environment can also affect a developing fetus's brain. After learning that pregnant women in Nairobi, Kenya, were not including protein in their diets in an attempt to reduce the size of their babies and have easier deliveries, she immediately wondered what effect this nutritional strategy had on the development of the brain. For the answer, she turned to her rats in the laboratory. When the rodents ingested a low-protein diet during pregnancy, the offspring were half the size of the normal-diet animals. What about their brains? When the protein-deprived rat offspring were housed in the classic enriched environment that Diamond made famous in the world of behavioral neuroscience, their brains did not respond with enhanced complexity. At least for the rats, quality protein in the diet of the pregnant female was necessary for the brains of her offspring to be able to respond appropriately to complex, challenging environments. But was it too late for these pups? Placing the rat moms and offspring on a high-protein rescue diet after delivery restored the brain's ability to respond to enrichment. These findings don't necessarily indicate that human children born to mothers who consumed low-protein diets during gestation are impaired; however, it is certainly plausible considering that neurons are essentially the same, regardless of the species.

In just the past century, food has transformed from a hard-earned resource to a cheap, widely available commodity that is virtually impossible to escape. Our brains are designed to approach and consume essential nutrients when they are encountered—as our ancestors were never quite certain when the next meal would be. This efficient, inflexible system presents many problems in today's society. The drastic changes in the rules of food consumption have no doubt had a critical effect on our brains and health. Informative rodent research may be our last hope to recover some of the damage we have so quickly inflicted on our species. Of course, we have been quite creative in the *inflicting damage on the health of our species* department. In the next section I'll address the hazards associated with the consumption of another type of substance—of the nonnutritive, psychoactive variety.

Rethinking Addiction: A View from Rat Park

Human universal healthcare discussions have been dominated by the most costly causes of illness. Along these lines one condition stands out. Whereas costs associated with obesity are outrageous (estimated to have reached close to $80 billion), the National Institute on Drug Abuse (NIDA) recently reported that substance abuse costs the United States approximately $484 billion each year. For comparison, diabetes and cancer cost our society $132 billion and $172 billion, respectively.

In their book *Prescription for a Healthy Nation*, Tom Farley and Deborah Cohen, both medical doctors, report that no drug comes close to nicotine's addictive lure. Recent statistics indicate that a whopping 20 percent of adults in the United States (61 million people) are current smokers. Most relevant to the healthcare system, the delivery system of this addictive substance accelerates death in many users, killing 440,000 Americans each year—equivalent of three 747s filled to maximum capacity crashing daily. This number represents twenty times more deaths than those caused by HIV, and this estimate doesn't include the approximately 50,000 Americans dying from secondary smoke. In a chilling statement, Farley and Cohen report: "If we cured all other forms of cancer, vanquished the AIDS epidemic, prevented all murders, suicides, and car/plane crashes, and wiped out all deaths from alcohol, we still would not have saved as many lives as we could by merely cutting smoking rates in half."

Thus, when thinking about a healthier country, it is difficult to escape the toxic effects of addiction, especially the threats posed by nicotine and cigarettes. Even though we know exactly what causes these negative health consequences—consumption of the drug—addiction persists in virtually every society. Why is this the case?

Rats have provided an immense amount of information about the case of addiction. When they are housed individually in their cages and allowed to self-administer a drug by pressing a lever, they increase consumption, showing evidence of withdrawal on the cessation of the drug. If given a preference test in which several bottles are presented with various liquids such as water, sugar water, and drug-laced water (water with morphine or cocaine, for example), rats will consume the drug-laced water at higher rates than the other solutions. When given injections of drugs in certain environments and asked if they prefer that environment or an environment in which a saline solution was administered, the rats prefer the drug chamber.

All of this evidence points to the observation that, similar to humans, rats become addicted to certain psychoactive drugs. These drugs hijack the brain's reward system and motivate the animal to seek and consume the drug, at the expense of other behaviors important for a healthy lifestyle. Indeed, an army of researchers have received funding from the NIDA to conduct research using rat models, in hopes of identifying the critical components for risk factors and clues about reducing consumption rates and diminishing the painful effects of withdrawal that so often lead to relapse. There's no denying that the rats have contributed valuable information about the mechanisms of addiction. As described in Chapter 3, the nucleus accumbens, the brain's pleasure center, as well as the neurochemical dopamine, have starring roles in this unfortunate healthcare saga.

But we may have been missing a critical point as we have tried to understand the addiction syndrome. Bruce Alexander, a researcher in Vancouver, British Columbia, and his colleagues at Simon Fraser University published a study in the early 1980s suggesting that the addiction story the rats are telling has been miscommunicated. Alexander was struck by the disconnect between the natural environments of rats, characterized by rich external and social stimulation, and the stark conditions of the laboratory cages, characterized by wire mesh, a water spigot, and a few pellets of prepared food. What if, the Canadian researchers wondered, rats were housed in an environment that was closer to their natural habitat and then given the option of consuming drugs?

The Simon Fraser psychology team conducted just that study. The habitat they created was about two hundred times larger than the standard rat cage, had walls with colorful murals painted on them as opposed to wire mesh, and contained lots of objects to interact with and climb on. Also important, rats didn't occupy the new environment in isolation. Groups of sixteen to twenty rats of both sexes were placed in the environment. The pictures Alexander shared with me are fabulous: The rats interact with their complex environment, hang out with each other, and take naps in hiding places, for example. The researchers dubbed this new stimulating environment "Rat Park."

Welcome to Rat Park

Subsequent investigations of the Rat Park habitat suggested that there was something about the combination of the social and enriched environment that offered protection against the emergence of addiction. The reams of rat studies using isolated rats were biased in an important way—the apparent pervasiveness of addiction in the rats was significantly influenced by their barren environments and impoverished social lives. If given the option of romping through an obstacle course, socializing with rat friends, or consuming morphine, they were less likely to choose the drugs than were animals with no meaningful options. These preliminary findings suggest that traditional rat studies may be most applicable to humans living in isolated, impoverished conditions. It is not surprising that even with all the drug screening, security checks, and lack of availability, higher addiction rates have been reported in prisoners than in nonincarcerated populations.

In my opinion, regardless of the lack of specificity in the Rat Park studies, the results fall in the blockbuster category. These data suggest that, unlike conclusions from the isolated, caged rat studies, there was more than a single cause of addiction. In addition to the drug itself, other factors were at play and this series of studies was the launching pad of a potentially revolutionary shift in addiction theory. But that revolution never happened. I have been a student of behavioral neuroscience for thirty years, and it wasn't until I was researching this chapter that I ran across this fascinating piece of addiction history. The Rat Park research has never been covered in any of the texts

I've used to teach behavioral neuroscience for the last quarter of a century, it has never been the subject of any of the hundreds of talks at psychological and neuroscience conferences I've attended, and it has never, to my memory, been part of the literature review of the many research articles I've read on the topic of addiction over the years. Why?

I was anxious to speak with Alexander, who is still thinking about these issues, to get his input on the research quandary. A delightful man, he was happy to discuss the little-cited research he championed three decades ago. Similar to my response to the findings, he thought these results were huge when he and his colleagues conducted the studies. It is not surprising that they submitted their initial results to the most influential science journals, *Nature* and *Science*. Both journals rejected the papers. Finally, the journal *Pharmacology, Biochemistry, and Behavior* published the results in 1981; this journal is well respected, but doesn't have the readership of the other two journals; thus this program of research was left essentially unnoticed from that time forward.

Aspects of the studies were replicated in the years after the initial Rat Park studies, but these have also failed to draw the fanfare generated by research championing the importance of genetic profiles associated with addiction vulnerability and the creation of new drugs to treat drug addiction (a treatment strategy that has always perplexed me).

Regardless of their impact, the Rat Park studies are informative about some of the lingering mysteries associated with addiction. For example, an interesting line of research related to human addiction reported that men enlisted in the army during the Vietnam War reported much higher rates of illicit drug consumption before leaving Vietnam than on their return to the United States. In one sample, it was estimated that 20 percent of the soldiers were addicted to heroin or opium; however, about a year after returning to the States, only 1 percent showed signs of opiate addiction. Was isolation from a supportive network of family and friends, coupled with the obvious stress of combat, a contributing factor to addiction in the soldiers? If so, these results emphasize the importance of lifestyle factors in addition to the power of the drug itself in establishing and maintaining addiction.

So the rats appear to be telling us that if we're interested in understanding the factors related to addiction susceptibility in human populations beyond prisoners, we need to consider addiction vulnerability in animals housed in natural, enriched social environments. After all, as significant as addiction is to the health of the United States, not everyone who uses a drug or takes a drink of alcohol develops an addiction. What underlies this important differential effect? Genetics are no doubt important, but there's certainly more at play here. Alexander's more recent work has been directed toward anthropology and a focus on humans who are disconnected from their social support systems. In his own hometown, Vancouver, he has postulated that high rates of addiction are due to the observation that, in his words, Vancouver is a "city of dislocation." From the time of the initial settlers from England in 1862, the city has forced Native Americans from their ancestral villages. Coupled with the fact that the European settlers were in turn dislocated from their own homelands and, with the completion of Canada's first transcontinental railway in 1886, the city, nicknamed "Terminal City," became a popular destination point for many migrants.

This social isolation, Alexander argues, has contributed to another nickname given to Vancouver—"Addiction City."

And what about that nicotine addiction, the leading avoidable cause of disease in the United States? Have the rats provided additional information about that health budget buster? George Koob and his colleagues at the Scripps Research Institute in La Jolla, California, have conducted research suggesting that nicotine addiction in rats involves the participation of a neural sidekick—that is, the brain's stress response. I know you're thinking it: Just how do they get the rats to smoke those little cigarettes? Well, there are some things rats just won't do, so the nicotine was either directly delivered into the jugular vein via a catheter or, in other phases of the study, the rats were allowed to press a bar so that the nicotine was delivered directly into the brain. Research has shown that corticotrophin-releasing hormone (CRH), a stress neurochemical working in the brain's most prominent fear center, the amygdala, produces a powerfully negative withdrawal state, making it very difficult for the smoker to continue in an abstinence phase.

In the study conducted by Koob's team, when a drug that blocked nicotine receptors was given to rats addicted to nicotine, they exhibited more anxiety-like behavior. Using a clever task, these researchers placed the rats in a cage with a prod extending into the interior that emitted a mild shock when the animal eventually touched it, then the shock was deactivated. However, this initial encounter with the shock prod typically makes the animal nervous and it starts digging in the cage's corncob bedding, shuffling the bedding in a heaping pile covering the shock prod. It's simple: out of sight, out of mind. Past research suggests that the quicker the animal starts burying and the longer it buries, the more anxious the animal. When the nicotine receptors were blocked in the addicted rats, they spent more time burying the shock prod, suggesting that being cut off from the effects of the drug exaggerated the brain's stress response.

Thus it is likely that the uncomfortable stress response plays a big role in driving the most well-intentioned person trying to kick the cigarette habit toward that pack of cigarettes. Consequently, stress management training may be a powerful treatment strategy for painful withdrawal episodes. And according to the rats, another effective treatment strategy for addiction is to engage in fulfilling lifestyles (perhaps Human Park?) that decrease the likelihood of consuming drugs in the first place.

The Rochester Rats Reveal the Immune System's Best-Kept Secret

Of course, Nature has its own universal healthcare system, the immune system. This system is as complicated as it is fascinating—consisting of immune organs, antibodies, killer cells, and messenger systems that send frequent reports about any pathogens brave enough to cross the heavily patrolled body boundaries. Up until the 1970s, students learned that the immune system was an independent contractor of sorts in the body. Immune Systems Inc. appeared to be completely self-sufficient, working in isolation of the body's many complex physiological systems. That is, until University of Rochester's psychologist Robert Ader and his immunologist collaborator

Nicholas Cohen made some rather odd observations in rats that would open the door to an entirely new discipline.

When Ader and Cohen were investigating the effects of an immunosuppressant drug on rat health, they cleverly designed a study to disguise the drug in coveted sweet water. The study seemed to progress uneventfully, until the scientists thought the study had ended. This is why it is beneficial to always have your eyes open when you're working with rats—you just never know when they are going to reveal an important evolutionary secret. The memorable event was that after the immunosuppressant drug was removed from the sweetened water, the animals got very ill, even died. It was as if just the memory of the drug-laced sweet water were enough to compromise the animals' immune systems. The immune system was suppressed when the drug in the sweet water entered the body. Even though the animal wasn't consciously aware that the drug was in the water, an association was made between the rat's immune functions and the sweet water—the stimulus that stood out to the rodents. Although Ader and Cohen knew they were seeing something revolutionary, they needed to design a new study to confirm it.

The study they designed provided strong support that the immune system had been conditioned to respond to a certain stimulus—that is, the sweet water, even after the drug was removed. The term *conditioned* is a code word for "learning." If the immune system actually learned from experience, that means that a translator of sorts would be required to communicate certain behavioral responses. There appeared to be only one logical candidate for this translator job—the nervous system. If the immune system could be conditioned, this would open up an incredibly exciting new array of possible ways to alter immune functions.

This seminal article was published in the journal *Psychosomatic Medicine* in 1975. Similar to the situation described for Alexandar after the Rat Park studies and Garcia after discovering conditioned taste aversion, Cohen and Ader waited for the respect and wonder of neuroscientists and immunologists across the world. However, scientists were more comfortable with the established notions of the immune and nervous systems working independently. Consequently, the results were met with a lot of skepticism, and those who tried to jump on this revolutionary research bandwagon had a tough time getting their work published. The immunology journals didn't appreciate being told that their ideas had been wrong all these years, and the neuroscience journals didn't seem to be interested in immune functions. So instead of basking in the glow of their research success, Cohen and Ader found themselves fighting for the creation of a new discipline, psychoneuroimmunology. The Psychoneuroimmunology Research Society was thus created, and their official journal provided a publication home for innovative research articles. Growth and acceptance occurred slowly and steadily; today I view the discipline of psychoneuroimmunology as one of the most exciting areas in neuroscience.

The field of psychoneuroimmunology, however, is still in its infancy, and scientists continue to explore the parameters of this research. Thanks to the rats, now that we know that the immune system is influenced by the nervous system, we understand how being exposed to chronic stress may compromise immune functions or how stress-reducing psychological tricks may enhance immune functions. Still, there are many questions yet to be answered. In Chapter 6, we'll learn more about how social interactions also influence immune functions.

A Final Dirty Little Secret

Most of us strive to keep our home, hands, clothes, and even the air, as clean as possible to ensure the healthiest immune functions. Rats, however, tell us that, in many cases, a little dirt doesn't hurt; in fact, it may be just what the doctor ordered. Duke University immune researcher William Parker and his colleagues recently set out to investigate a hunch called the *hygiene hypothesis*. This hypothesis suggests that living in a very clean environment, typical of cultures with modern medical practices, may be underwhelming for the immune system, leading it to occasionally act out in the form of allergies and other autoimmune diseases. Allergies, for example, are more prevalent in cultures with modern medical practices than in cultures with less modern practices. Parker's group assessed antibody production in rats living in both wild and laboratory conditions. The laboratory rats could be viewed as inhabiting a culture with modern medical practices whereas the wild rat habitats represented less modern cultures.

Although one might expect the removal of germs from one's environment to contribute to the healthiest immune systems, the wild, dirty rats possessed more stimulated, protective, less pathogenic responses. Thus these results suggest that clean, sterile environments may not be challenging enough for the immune system, potentially explaining the high rates of allergies we see today. Of course, with other medical challenges, such as flu, clean environments (especially clean hands) positively influence health. The wild rats, however, suggest that, for some immune functions, a cleaner environment may be too mundane for a system designed to work so thoroughly to protect its valued host.

As we continue to learn the complex secrets of the immune system we will be better prepared to handle viruses and bacteria that penetrate our immune boundaries. In addition to identifying the ways to keep our bodies free of illness and disease, a thorough universal healthcare plan should be vigilant about mental health. Consequently, the idea of building emotional resilience as a means of building protection against the emergence of mental illnesses has also become an exciting new area of rodent research (and one of my personal favorites!), one that we will turn to in the next chapter.

CHAPTER 5

|||||||||||||||||||||||||||||||||||||||||||||||||||||||||||||||||||||||||||||||||||||

Enhancing Emotional Resilience

Several years ago, my colleague Craig Kinsley and I wrote a textbook about the neurobiological factors associated with various mental illnesses. After conducting extensive reviews of the biomedical literature, it was clear that stress was the most dangerous threat to our mental health. Although genetic and other physiological factors are involved, a stressful incident is often associated with the first expression of schizophrenia symptoms, chronic stress is clearly associated with the onset of depression, and stressful situations prime the pump for the emergence of symptoms of various anxiety disorders such as phobias and post-traumatic stress disorder. On the other hand, there's no doubt that our stress responses serve a valuable survival function—enabling us to muster up the energy to jump out of the way when a car is about to hit us. But stressors found in today's world have changed since the time of our earliest human ancestors. Contemporary stressors rarely require us to run from a lion or fight for our food; we can typically ride out our stressful situations without leaving our office chairs, yet the physiological effects are the same.

The stress response is complex and appears to live a double life. It is simultaneously viewed as being both life-saving and life-threatening. It accompanies our most pleasurable responses and is responsible for our lowest lows. I've been investigating the effects of stress on behavior, brains, and bodies since my early days in graduate school, and I'm still just as intrigued by this topic as ever before. As indicated by the title of this chapter, my focus has shifted from mapping out the deleterious effects of stress to learning how to build emotional buffers, a form of stress busters, protecting us from the harmful effects of chronic stress. Consequently, *resilience* is the buzzword in my lab these days.

Rats have contributed much toward this topic, and the elusive pieces of the stress puzzle are finally starting to make sense. With this information, it is becoming clear that the effects of stress can be harnessed and tamed to keep wild responses from running rampant through our emotional lives. The most informative piece of the puzzle relates to coping strategies, a topic emphasized in this chapter. Rats and humans alike experience stress throughout their lives, and it is becoming increasingly clear that the way stressors are perceived determine how lethal the dose will be. Educating the public about stress and coping strategies should also be an essential component of any universal healthcare plan (discussed in the last chapter). In the United States, costs associated with stress-related illnesses exceed two hundred billion dollars each year. Stress leads to cardiovascular disease, diminished immune functions, depression and anxiety disorders, diabetes, and gastrointestinal illnesses, and the list goes on and on, as do the costs.

Consider a form of stress we can all relate to: having a scary dental procedure. Did you know that researchers have found that your dentist can relieve the intensity and duration of pain experienced during various dental procedures simply by taking time to give you information about what he or she is going to do before bringing out the needles and drills? According to these findings, your dentist's speaking skills may be more important than his or her surgical skills. In this case ignorance is *not* bliss. If your dentist ventures into your mouth with no explanations, your brain's hippocampal network goes crazy, plotting out every conceivable worst-case scenario and accompanying escape plan should the pain reach an intolerable level. All of this brain activity exacerbates the pain perception to make sure your neural networks are on the case. On the other hand, if everything is going according to the plans described by your dentist, even if the procedure is painful, your hippocampal problem-solving network is kept at bay, and your pain is more tolerable.

Why this foray into dental-pain management? First, it emphasizes the importance of context in the determination of stress responses. Further, because I consider the topic of stress to be so profoundly important, I just can't help but lapse into a mini-lecture about the origins of stress research and, when I get to the section describing our recent work that reveals valuable secrets about building resilience, a certain degree of specificity will be required to convey the relevance of our findings. The complex nature of stress doesn't allow for watered-down descriptions—I think too highly of my readers to discuss one of our brain's most fascinating response systems in sound bites. Thus, similar to the accommodating dentists who issue fair warnings about upcoming procedures to minimize distress, I'm issuing a bit of a warning that a few intense descriptions of the complex stress literature lie ahead. But I promise, any pain associated with an occasional academic detour will be minimal and short-lived and, unless you find yourself in one of my classes in the future, there will be no test! Don't worry, there are plenty of entertaining and informative rat tales, ranging from stories of bold diving rats to spoiled-rotten rats, to inform us about the most effective coping strategies. We'll begin our stress story with a Canadian researcher who had a problem holding on to his rats; no doubt a less vigilant researcher would have let this line of research slip through his hands!

Straining to Understand Stress

One would be hard-pressed to find a medical researcher who was more influential and prolific than Hans Selye. Working at McGill University in the 1940s, Selye serendipitously discovered a cascade of physiological events that typically accompany challenging situations. He began his research journey, however, on a different path. Selye was excited about assessing the effects of a specific hormone extract on rats. He studiously injected each rat and recorded the physiological effects. He could hardly contain himself after running the experimental group—that is, after observing the group that received the actual hormone extract. The animals were obviously sick. They had swollen adrenal glands, shriveled immune glands (thymus glands, to be exact), and ulcers in their stomachs. He felt he was on the verge of a medical breakthrough. His hopes were later squashed when he ran the nondrug control group and found exactly the same result. How could an injection of saline, the physiological equivalent of nothing, have a harmful effect on the body? What was going on in this crazy study?

These rat findings contained extremely important clues about the stress response; would Selye be able to crack the code and decipher the results? To his credit, he closely observed the rats, pondered the data, and finally arrived at a reasonable explanation for the confusing results. If he had to be honest with himself, he wasn't the most competent rat handler. While he was injecting them, he sometimes squeezed a little too tightly. Owing to his anxiety, he sometimes dropped the rats, which meant he had to chase them around the room with a broom so he could finish the experiment. Thus the rats in this study were getting more than a dose of a hormonal extract—they were getting a pretty heavy dose of terror as well. The control animals endured the same unpleasant circumstances as did the experimental animals, explaining why, regardless of the type of injection, all the rats exhibited similar symptoms.

As he considered these intriguing results, Selye saw the importance of the noxious stimuli and, against the advice of his mentor, decided to drop his interest in the hormonal extract. At that point, he went to the proverbial research drawing board and designed additional studies focusing on the effects of noxious or adverse experiences. If he placed the rats in cold, hot, loud, or unstable conditions, he noted a similar pattern of results—the same triad of responses his injected animals developed. Selye was documenting quantifiable effects of adversity, a very different endeavor from the more traditional medical research of the time. This research was so bizarre that he didn't even know how to frame it or what to call it. He bypassed naming this new concept in his first publication on this topic, which appeared in the prestigious journal *Nature*. His article, titled "A Syndrome Produced by Diverse Nocuous Agents," provided a brief introduction to what Selye viewed as the most basic form of reaction to various threatening situations, a response he referred to as the *general adaptation syndrome.*

In search of a descriptive term to use in his writings, Selye went to the engineering literature, where he was intrigued by the notion that strong materials became vulnerable when they were worn down. In this vulnerable state, when the next pressure came along, the material was more likely to snap. Selye saw similarities with his rat observations. Although he was actually reading about a process called *strain*, it was discussed in the context of a related term, *stress.* Most scholars in this area agree that Selye meant to name this new response *strain*, but being born in Hungary, Selye misunderstood the English-language descriptions and chose the word *stress.*

The term *stress* subsequently generated confusion as scientists looked to the engineering analogy and scratched their heads. Was stress the cause or effect of the response Selye had described? Selye used the term *stressor* to refer more specifically to the stimulus causing the damage; however, more accurate terminology would have been stress as the cause and strain as the effect, or the body's response. Nonetheless, Selye went on to describe his general adaptation syndrome in which he emphasized the consistency of the stress response. He also emphasized the importance of a hormonal cascade, focusing on the critical role of the stress hormone corticosterone.

Although Selye's contributions to this field are extremely valuable, there are certainly aspects of his theories that have been modified and adjusted through the years. Most important for this chapter is the observation that stress responses can be variable as opposed to always being fixed or constant. Biological predispositions and life experiences can modify the body's responses to stressors. Speaking for myself, jumping out of an airplane, even with a parachute, would no doubt kick in an extreme stress response that would take quite a while to totally recover from; however, my postdoctoral assistant, Catherine Franssen, chose skydiving as her favorite way to cope with the stress associated with graduate school. Two people, two very different responses to the same stimulus. As you are probably thinking, at some level, stress responses are more predictable. That is indeed the case; in fact, responses related to imminent danger (being thrown off a building) in which we have little control are high on this list.

Peter Sterling, a neuroscientist at the University of Pennsylvania School of Medicine, recently contributed to the evolution of the stress concept by questioning the value of another health-related term: *homeostasis*. You probably remember this term, related to the balance of physiological systems in your body, from your sixth-grade science class. Nineteenth-century physiologist Claude Bernard introduced the idea to emphasize the importance of preserving a constant, consistent internal environment in maintaining a healthy system. Harvard physiologist Walter Cannon, however, used the term *homeostasis* in the

1930s to refer to the body's optimal stable, consistent internal environment.

Sterling and others have offered a convincing argument that, for the most part, healthy responding is not a result of maintaining constancy; quite the contrary, it is a result of maintaining optimal responses by *changing* our systems in appropriate ways. If you are running a marathon, you will not live through the experience if your cardiovascular system does not change in various phases of the run to accommodate changing cardiovascular demands. After you eat your favorite dessert, your health depends on your body's ability to change the production of insulin to break down the sugar load you've introduced into your system. And when you see a child run into the traffic, your stress hormones need to be released at a higher level to give you the mental and physical energy to navigate the traffic and scoop up the child. Healthy responding is not about maintaining constancy; it is about being able to change our responses to meet the demands of the environment. Sterling and others have suggested that *allostasis*, defined as "maintaining stability through change," is the term that should replace the old-fashioned notion of homeostasis. Revisiting the engineering analogy one more time, Rockefeller Institute neuroscientist Bruce McEwen introduced the term *allostatic load* to refer to the strain placed on the brain and body after stressful experiences should necessary changes fail to occur in a timely manner. If it takes a long time for the cardiovascular system to return to baseline conditions every time you engage in a demanding task, such as dragging heavy luggage around the airport, allostatic load could eventually lead to cardiovascular disease.

Thus the toxic effects of stress, arguably one of the most important biomedical discoveries of the twentieth century, were initially detected in laboratory rats by several enlightened scientists willing to rethink the scientific dogma related to the body's checks and balances. What exactly do the rodents tell us about this complex stress response? McEwen's Rockefeller Institute rats have revealed valuable information about the effects of allostatic load on the brain.

The Incredible Shrinking Neurons

McEwen's experiment was straightforward. Inject the stress hormone corticosterone into the rats for three weeks and then assess the size of the neurons in the brain's hippocampus, a brain area involved in the critical task of spatial learning in these animals. Why look in this brain area? Because the hippocampus is rich in glucocorticoid receptors, it's the first place curious neuroscientists look for stress-related brain changes. Perhaps Nature chose an area involved in learning and memory for this neurochemical's home base since it would certainly be beneficial to remember the dangerous stressors in the world around us (including traumatic visits to the dentist described in the chapter opening). In many cases, one-trial learning is essential to staying alive. Nature has little tolerance of animals requiring several trials of walking up to a lion, or cliff, or alpha male before learning that it may not be a smart move.

McEwen's team patiently traced the entire distance covered by the many thread-like processes of the neurons in the hippocampus. What did they find? These neurochemicals had definitely left their stamp on these neurons. There were fewer branching points and shorter processes. Could this neuronal version of downsizing affect the animals' memories? Ability to learn new tasks? Assessment of future stressful situations?

McEwen's next step was to determine if actual psychological stress would wreak the same havoc on the hippocampal nerve cells. Instead of injecting stress hormones in the animals, this time he exposed the rats to an uncontrollable stressor—they were placed in a Plexiglas restraint tube for a couple of hours each day for about three weeks. The animals weren't physically harmed in any way, they were just exposed to the psychological stress of not being able to escape, similar to being stuck in rush hour traffic—neither rats nor humans like these situations. The restrained brains looked very similar to the hippocampal cells in the rats that had been injected with stress hormones. Regardless of the source, stress hormones delivered in a chronic fashion over the course of three weeks weakened the brain's cells in critical areas.

Three weeks is a relatively long time. Just how much stress is necessary to shrink our brains? My students and I used another stress procedure, with the rats housed in cages with adjoining running wheels, to address this question. When their feeding schedule is manipulated so that they can eat for only one one-hour session per day, the rats' behavior becomes, well, odd. Even though food is restricted, they run more and more and show all the symptoms of Selye's original stress syndrome: swollen adrenal glands, suppressed thymus glands, and lots of stress ulcers in their stomachs. After just five days of this stress, we saw similar shrinking effects in their hippocampal neurons.

Before you run to the nearest source of formaldehyde in an

attempt to preserve the few neurons you think you have left, there is some good news, so hang in there. First, there is little evidence suggesting that these nerve cells are actually dying, at least in the short run. The damage, or restructuring, appears to be reversible; once the stress is relieved, the neuronal processes will likely start to blossom once again. Although shrinking neurons sound pretty scary, this may represent an efficient energy-saving plan. Because the brain is such an energy hog, using up to 20 percent of the body's resources at any given time, some neuronal downsizing may be necessary for short-term survival in stressful situations. The overactivation of the stress hormone axis may also require a little downsizing to avoid system overload. Once the costly threat is gone, the neurons can once again operate at full capacity. Second, according to some smart rats, this scary neuronal downsizing may be avoided altogether if the stressful situation is initially approached in certain adaptive ways.

A Tale of Two Rats

Several years ago I was struck by a very powerful picture of two rats, presented by neuroscientist Sonia Cavigelli, currently at the Pennsylvania State University. Cavigelli described the effects of varying responses to the mild stress of novelty (simply being exposed to a new object) in rats. In her study, conducted with Martha McClintock, at the University of Chicago, young rats were profiled as neophilic (bold, novelty seeking) or neophobic (shy, uninitiated). How was this determined? Rats were individually placed in an open arena, something rather scary for a rat. A new object was placed in the center of this arena, and Cavigelli and McClintock simply recorded the time it took for the rats to contact the novel stimulus and the amount of time it spent investigating the object. Using brothers within the same litter, practically genetic clones in inbred laboratory rat strains, two very diverse patterns emerged: Some rats avoided the stimulus and clung to the walls of the arena, whereas others investigated the new objects. The scientists had identified two rat personalities: shy/neophobic and bold/neophilic.

When the rats' stress hormones were assessed, the bold rats had a quicker recovery, meaning that their levels returned to baseline faster than their shy counterparts—translating into a lighter allostatic load than found in the shy rats. The animals were assessed every six months, and the characteristic profiles persisted. Confirming that these rat personalities were long lasting, the shy group remained shy and the bold group remained bold, with the initial stress hormone effects persisting. Remarkably, after allowing the rats to live out their lives in the laboratory, it was documented that the bold rats lived longer; in fact, the shy rats' lives were about 20 percent shorter than the bold rats.

And what about the rat picture I mentioned earlier that stopped me in my tracks? Two rats were featured in the picture, one looking spry and chipper, the other hunched over, looking very old and tired. These rats were brothers from the same litter, exactly the same age, the younger-looking animal was profiled as a bold rat when it was young and the old-looking animal was profiled as a shy rat. Of course, this bold response may not be as healthy in the real world; I often tell my students that the bold rat probably would have died in a motorcycle accident in the outside world. All things being equal, however, you're better off boldly going where no rat has gone before. At least that's the case in the lab. In a subsequent study focusing on females who spontaneously developed mammary tumors, the shy rats died about six months

> You're better off boldly going where no rat has gone before.

earlier than the bold rats. When your life span is only about two years, six months is a significant amount of time, the equivalent of adding fifteen or twenty years to our life span.

So research suggests that stress responses are *not* created equal; in this case, having a temperament that is associated with an enhanced sense of exploration is more adaptive than being too shy to explore environmental surroundings. A key lesson in this research is that our day-to-day stress responses—how we respond to a new coworker, software program, or traffic pattern— have a significant impact on our well-being and longevity. It is quite clear that our stress responses need to kick in when we're faced with life's big stressors, but individual coping strategies may determine the efficiency of our more mundane responses to life's daily hassles. It is becoming increasingly clear that any information about effective coping strategies can have powerful health benefits. Have the rats revealed other adaptive coping strategies?

Adaptive Coping: The Slinky Strategy

When Richard James was working as a mechanical engineer at a naval shipyard in Philadelphia in 1943, he saw something that would change his life. A torsion spring fell from a table and hit the floor with an odd wiggling motion. That night he told his wife that he thought he could make a toy out of the spring. After working on springs for two years, he generated one that had just the right amount of elasticity and tension, allowing it to walk down steps. His wife looked for a name for this new toy. When she went to the dictionary she found the word *slinky*, defined as "sinuous, stealthy and graceful of movement." Since the 1960s, the Slinky has been a staple among affordable and imaginative toys.

What does the Slinky toy have to contribute to a discussion on adaptive coping strategies? The fascinating part of the Slinky is its "smart" ability to adapt to the environment when determining its movement. Placed on top of a tall step, it stretches down to the next step and continues with this routine until hitting a different type of landscape. Responding to a decline as opposed to a step, the Slinky adopts a less expansive step. At the end of the journey, the Slinky recoils to its conserved state, ready to respond the next time the toy's owner sets it in motion. And, perhaps most impressive, when stretched to its maximum length, the Slinky snaps right back into shape with no apparent wear or tear once the pressure is released.

Other toys present during the 1960s were very different. Wind-up toys continued with the same movement regardless of changes in the environment. If a wind-up soldier hit a wall, it would continue to march into the wall until it needed rewinding. The movement of pull toys and remote-control toys could be altered to meet the changing demands of certain terrains, but the control was from an external source. The Slinky had it all—the ability to use intrinsic mechanisms to respond to changes in the environment. It may not have always been perfect, but its self-regulated flexibility was impressive. Of course, mammals are more complex than Slinkys,

The Slinky had it all.

but I'm convinced there's a valuable lesson to be learned from these toys, a lesson that emerged from both the scientific literature and the rats in my lab. Let's start with the literature.

After reading about the shy and bold temperaments in rats, I was determined to learn more about the most adaptive coping strategies in mammals. My students and I were initially intrigued by some unique studies conducted on Dutch pigs. If piglets were held on their back for just a minute (a procedure known as the *back test*), researchers observed that some would struggle and wiggle to escape, whereas others would remain still while the time passed. The active animals were referred to as proactive, and the more passive pigs were categorized as reactive. The active animals, perhaps similar to the wind-up toys, seemed to be more rigid in their responses, always responding the same way, whereas the reactive animals were viewed as more variable copers, sometimes responding, sometimes not. I wasn't quite sure, however, if the animals the Dutch scientists were describing as reactive were of the more passive pull-toy variety or the slicker, adaptive Slinky variety.

The story got more interesting when the researchers tested the pigs a second time and mentioned an annoying observation. Some of the pigs changed coping categories—that is, pigs profiled as reactives became proactive, and proactives became reactives. This wishy-washy group was viewed as "doubtful" by the researchers and tossed aside in some studies. We started to think that the reactive, variable copers may represent two groups, one that was consistently more passive (shy, perhaps) and one that was consistently more flexible. We had to know more about these coping strategies, but there was absolutely no way my provost was going to allow me to conduct pig research on our small Randolph-Macon campus. In true Slinky fashion, I had to be flexible, so I adapted the back test to rats so I could answer my questions, and I was hopeful the rats would cooperate.

My undergraduate student Kelly Tu was the first student to work with me on the back test research. We gently held just-weaned rats (about twenty-two days old) on their backs and simply counted the number of wiggles. The pigs commonly gave zero to three wiggles, but the rats gave us more responses, from zero to about fourteen wiggles. We categorized the rats with the highest responses as active and those with the lowest as passive and waited a week to test them again. Using the same format, we noticed that some rats were extremely active on both tests, some were extremely passive on both tests, and there was a group that switched categories on the second test. The direction of the switch didn't matter to us at the time; we were interested in the rats that displayed the most dramatic behavioral changes—they became our variable, or flexible, copers. Could it be that the flexible copers learned something about the first back test and changed their behavior accordingly in the second test? It seems that this would be a very adaptive, Slinky type of approach to life's changing demands. We wondered if the flexible copers had advantages over the two consistent coping groups. We had our work cut out for us.

Initially, we profiled the three coping groups and introduced the rats to the activity–stress paradigm described earlier in the chapter, where rats are housed in cages with activity wheels and are given a restricted diet. In this study, the rats could eat as much as they wanted for two hours per day, could run in the running wheel as they pleased, or could rest in the adjoining cage. True to their name, the active copers ran more than the passive animals, and, in this case, the flexible animals also ran more than the passive animals. During the midst of the paradigm, we collected fecal samples, or poop samples as we began to refer to them in the lab, to assess the stress hormone corticosterone. If energy expenditure was associated with the degree of stress, the active and flexible animals should have the highest levels of this hormone. That wasn't the case. The active animals did indeed have the highest level but the flexible animals had the lowest levels of stress hormones—in fact, the flexible rats had about half the level of the passive animals and 40 percent of the active animals. We were amazed with these initial findings. Even though the flexible rats were running as much as the actives, they didn't seem to be as stressed. Did these animals see their situations, perhaps due to their perceived sense of control, differently from the other groups?

We conducted other simple tests. For one, I used small hair clips like my daughters use (the ones that are clam shaped and open up when you pinch one end) on the rats' tails to simulate a bug bite. After we placed the clips gently onto the rats' tails, the flexible animals persisted longer with their attempts to remove them than the other coping groups. Further, the flexible animals interacted with a novel stimulus more often than the passive and active rats, even though there was no statistically significant difference between the flexible and the active animals. It was becoming clear that, if given a choice, it was best to be a flexible rat.

At that point we hadn't explored underlying brain mechanisms associated with the flexible, Slinky coping advantage, and we were eager to do just that. Another student, Darby Fleming Hawley, spearheaded that effort. She profiled the three coping groups and, this time, exposed them to a different type of stress, the kind that seemed to be most toxic to our brains—stress of the chronic and unpredictable variety. As I write this chapter, the economy has created this type of stress for so many individuals in the United States—loss of jobs, prolonged unemployment, and little control in regard to changing the situation. Such stressful situations translate into increased stress, strain, and allostatic load; regardless of what you call it, it's bad.

Instead of stressors such as unemployment and unexpected bills, we introduced annoying little things to the rats such as vinegar in their water, predator odors (fox urine is a favorite in the lab), tilted cages, and strobe lights. The stressors and schedules varied each day, similar to real life.

The most interesting behavioral response in this study was associated with one of the most popular behavioral tests of depression in rats, the forced swim test. When rats are placed in a tank of water they generally have two options: swim or float. Scientists have traditionally viewed the floating response as the weak or maladaptive response, sometimes called behavioral despair. The observation that antidepressants reduce floating and increase swimming in this task has added more credibility to the notion that swimming, not floating, in the forced swim task is the healthy response.

Given our doubts about the value of the swim task, we tweaked it a bit. We increased the size of the swim tank (typically a small cylinder) to the size of a small aquarium so the animals could actually swim. We weren't certain what the smartest response would be in an initial swim test but if the animals were indeed responding to the environment, the rats should adopt a floating strategy once they had learned that there was no way out of the swim tank. As we predicted, the rats' responses got very interesting in the second swim. The flexible copers, who had spent the least amount of time floating in the first test, switched to the most amount of time floating in the second swim, increasing their float time by 400 percent from the first swim. The passive and active copers increased their float time as well, just not as much. By the third swim, however, all three groups were on board and floating the same amount; they had apparently learned that there was no escape route and that they would eventually be removed—no reason to keep swimming in a panicked frenzy. Far from behavioral despair, the rats seemed to be telling us that, once it was established that there was no escape route, floating was an adaptive, energy-conserving strategy. This was an impressive demonstration of the rodent version of taking a behavioral chill pill.

> Floating was an impressive demonstration of the rodent version of taking a behavioral chill pill.

Our primary goal of this study was to investigate the brain

mechanisms associated with flexible coping. Could we identify any differences among the brains of the three groups? We focused on a prevalent neurochemical in the brain, neuropeptide Y (NPY), a substance that had been associated with resilience, as we discussed in Chapter 3. Past research indicated that the infusion of this substance into certain areas of rodent brains caused the animals to appear less stressed and anxious. As mentioned earlier, special forces soldiers, known as some of the most resilient among soldiers, have higher plasma levels of NPY than other soldiers, suggesting this chemical is also associated with resilience in humans. We looked in three brain areas involved in stress, fear, and anxiety and found that the flexible animals had higher NPY levels in two of these areas. I found these results nothing short of amazing: We identified a coping group that, without pharmacological supplements or even extensive training or therapy, had higher levels of a neurochemical involved in emotional resilience. Wow.

The stress hormone story was very different in this study. Recall that our first study used a consistent, predicable form of stress: Food was presented at the same time each day and the animals had 24/7 access to the running wheel. Every day was just like the previous day. In the chronic unpredictable stress study, however, everything kept changing; day by day, the animals never knew what to expect. In this case, the flexible copers had the highest levels of the stress hormone corticosterone, suggesting that they were in a constant state of readiness to respond to the next challenge. Thus the responsiveness of the stress response changed with varying types of stressors and stressor schedules.

The next study we needed to do was obvious. Could we use

effort-driven reward training, the training that required rats to dig up Froot Loops rewards (described earlier) and that built a sense of control by strengthening associations between effort and positive consequences, to make flexible copers out of active and passive copers? To find out, we designed a study to determine if we could help the vulnerable rats with a little behavioral therapy.

Spoiled-Rotten Rats: The Computational Source of the Stench

My student Alexandra Rhone led the next study. After profiling the rats' coping strategies, she assigned the animals to an effort-driven reward group (*contingent training*, in the sense that the reward was contingent on effort) or a control trust fund group (*noncontingent training*, as described in Chapter 3).

In my mind, our past findings and the existing literature in the area suggested that the interesting group in this study would be the contingent trained experimental group—the effort-driven reward group that had to dig for their Froot Loops. This enhanced sense of control over the environment would, I thought, be enough to bring the passive and actives around, increasing their NPY levels and modifying their behavior in adaptive ways. A rat version of behavioral therapy! The noncontingent trust fund group was the necessary control group to simply show that, without the training, no effects would be apparent. As you'll see, the rats revealed a very different story.

After four weeks of training, we conducted the cat ball persistence task (see Chapter 3) to determine if our trained rats, regardless of coping strategy, would still work harder to retrieve the elusive Froot Loops reward located inside the cat ball toy. They did; and the flexible animals also persisted longer, providing more evidence that these animals were some type of emotional superheroes. Next we conducted the swim task, with three installments, so it could be compared to the initial study. In the trained effort-driven reward group, the flexible rats remained the highest achievers—demonstrating more floating across all sessions. Once again, it appeared that they were quick to discern that there was no escape route from the swim tank. The training, however, failed to help the active and passive coping groups as we had hoped to see. In retrospect that made perfect sense: The animals that seemed to possess a predisposition for learning about changing environmental demands demonstrated faster learning about a changing environment (that is, after their behavioral training). In essence, we had built on their strength!

We were in for a bigger surprise when we turned to the trust fund groups. As described earlier, we fully expected this to be the boring control group in which the benign treatment of being given Froot Loops every day in the absence of any effort had to be assessed simply as a part of good experimental design. Just like the saline injections Selye used to confirm the negative effects of his hormonal extract, we viewed the trust fund group as the behavioral equivalent of nothing. But as Selye observed, the apparent control group held the most interesting clues. It was clear that something was different about the flexible copers in the trust fund group—they no longer learned to alter their float times across the three swims. All three groups exhibited relatively low float times in this condition; no one was learning from their past experiences and changing their behavior accordingly. What had changed for these potentially overachieving rats?

The rats in this study were telling us that their behavioral strategies were calibrated in a world that required them to expend energy for life's most valuable treasures—Froot Loops in this case—and when that formula was disrupted, their emotional resilience was also thrown off track. The daily presentation of the rewards, which required no extra effort and appeared as if they were just dropped from the sky, compromised their established effort-response probability computations or, in other words, diminished their sense of control. Thus the flexible rats lost their advantage in the cushy new trust fund world. So that was strike two in the hypothesis department for our study: Instead of having no effect on the coping responses, the trust fund condition erased the advantage typically shown by the flexible copers. The lack of a predictable contingency formula accompanying the presentation of life's sweetest rewards reset the behavioral computations underlying the rats' motivation to work for their rewards. They were now characterized by less flexibility in their responses and a shorter tolerance for work that didn't immediately produce a reward. Had we systematically spoiled our rats? Once again, animals that were more sensitive to associations between effort and consequences would likely be even more affected by the trust fund noncontingency condition; after the fact, it all made so much sense.

It is interesting that the trained flexible rats also had higher levels of a hormone associated with resilience, DHEA. DHEA, as mentioned in Chapter 3, appears to buffer against some of the toxic effects of stress hormones. Focusing on human resilience, military combat divers who successfully navigate challenging underwater diving courses (perhaps due to their enhanced sense of control gained by strong associations between their actions and outcomes that are learned during military training) also have high levels of DHEA.

From the results of our rat study, it was clear that the practice of dispensing unearned rewards disrupts the brain's computational formulas for work and rewards, compromising emotional resilience. With this revelation, my scientist persona faded to my maternal persona. As a mother, one of the most important lessons I feel that I need to teach my children is that positive consequences follow hard work. I constantly tell them that if they pick themselves up after each failure and keep plugging away toward their goals, good things will eventually come their way. As a parent, I wonder every day if I'm giving too many rewards to my daughters and if I'm providing enough opportunities for them to dig up their own Froot Loops. I take my lessons from my rat friends very seriously; they have no book to hype or talk show to promote; as long as I'm willing to listen to their messages, they will reveal important aspects of Nature's truth.

> The practice of dispensing unearned rewards disrupts the brain's computational formulas for work and rewards, compromising emotional resilience.

But could we be stretching these trust fund rat results too far? Are there documented similarities between the rat and human brains when it comes to motivation and decision making? University of Cambridge neuroscientist Trevor Robbins and his colleagues have drawn fascinating parallels between the impressive executive functioning areas of the cerebral cortex in rats and humans. According to their research, the rat cortex is a valuable model for understanding decision making in humans, providing strong evidence that our rat resilience models apply to human emotions. So, if we're guilty of merely giving our children rewards that aren't contingent on any type of work, we may be recalibrating their brains' behavioral probability formulas or, put another way, spoiling them rotten. These results reinforce the importance of teaching children that outcomes do indeed follow actions. When we do this successfully, our children will likely be more motivated to engage in the appropriate behaviors to earn their rewards. This valuable contingency is weakened, however, if a privilege is presented in the absence of these behaviors. Just like the rats, children are smart: Why waste the energy if it isn't always necessary? Of course, we can never create a world where good things always happen in a predictable fashion, but my guess is we can do a better job of constructing more accurate associations between effort and rewards.

Another little gem of a finding in this study related to something else the rats did when they were placed in the larger swim tank. The rats generally had two choices in this test: float or swim. As usual, I underestimated their behavioral potential because the rats in the study offered another behavior, this time gaining major style points! Some of the rats held their breath and dove to the bottom of the tank demonstrating beautiful swim strokes as they explored the tank, inspecting the corners for escape routes. How did they do that? The rats had never been in water before; they had never had a swimming lesson or watched underwater exploration on television. Something just told them to go under. Can you guess which animals spent the most time diving? That's right, the contingent-trained, effort-driven reward animals. They had learned that their actions were associated with positive outcomes and subsequently engaged in this bold response. I don't know if I'm daring enough to say they were more optimistic about finding an escape route, but this behavior may be considered an animal model of just such a wonderfully complex response in humans. As predicted, the dives diminished in the second swim once they learned there weren't any escape routes. But we all enjoyed this rodent underwater show while it lasted.

The scientific literature has yet to show us that winning a lottery, inheriting a trust fund, or stepping into the family business without even having to interview for the position leads to fulfilled satisfaction and emotional health. Once the noncontingent rewards outnumber contingent outcomes, perhaps the technical definition of spoiling our children, we confuse our children. Their levels of frustration and stress increase as they realize they no longer know how to respond to get what they want or need. This confusion sometimes results in a good old-fashioned temper tantrum, perhaps the least effective coping strategy in existence (but that depends on the parents' tolerance for high-pitched squeals!). I haven't been able to create temper tantrums in the rats . . . yet.

Superheroes, Emotional Regulation, and the Smarts to Pull It Off

I'm fascinated by superhero stories in our culture. Each superhero has a unique superhuman ability that gives him or her an advantage over mere humans . . . they can fly, distort their bodies, climb walls by generating a web, exhibit unbelievable strength, or detect sensory information that no one else can decipher. If you think about it, though, many of these characters share a common ability, the ability to remain cool under pressure. Stanford psychologist James Gross refers to this impressive ability as *emotional regulation*, an ability that may be enhanced by one's tendency to reappraise situations, such as reassessing the impact of being held on your back or swimming in a tank after an initial exposure (if you're a rat) or remembering that the last time you spent the weekend with your in-laws it wasn't that bad (if you're a human).

Whereas mere mortal humans may scream when falling to their deaths, these fictional superhuman characters, fortified by their sense of control thanks to their special abilities, are just as likely to throw out a clever one-liner, making light of the doomsday situation and knowing exactly what to do to save the day. It takes a very active executive functioning cortical center to hold the crazy, panicking emotional parts of the brain at bay—impressive stuff. Emotional regulation is not only found in fictional stories about human responses but is a very real part of everyone's emotional lives. Poker players display an impressive sense of emotional regulation with their poker faces. These emotional overachievers have the ability to restrain their facial expression even if they are about to go bankrupt. Emotional turmoil triggered by a horrible or fantastic hand appears to be effortlessly suppressed.

In many situations this regulatory control works to our advantage, especially when we're not in immediate danger. In daily challenges that simply require us to make the most informed decision, possessing both an impressive degree of emotional regulation and a heightened sense of control over our environment allows our cortical brains to step back from the effects of the stress hormones and neural panic buttons to accurately assess the situation, generate the most accurate probabilities of certain outcomes, and make the best decisions.

Although amazing strength and x-ray vision most assuredly lead to enhanced survival, the calculating cerebral cortex is perhaps the most impressive survival weapon possessed by mammals. Kind of a cognitive and emotional *shazam!* With this realization comes the responsibility of maintaining the brain's emotional health by keeping the action-to-outcome ratios real—a premise that we'll revisit in Chapter 12. But we're getting ahead of ourselves; in the next chapter we'll discuss another one of the rat's secrets for reducing life's burdens: the therapeutic value of simply reaching out and touching someone.

> The calculating cerebral cortex is perhaps the most impressive survival weapon. . . . Kind of a cognitive and emotional *shazam!*

||||||||||||||||||||||||||||||||||||||||||||||||||||||||||||||||||||||

CHAPTER 6

The Value of Social Diplomacy

I'm guessing you didn't conjure up images of rats when you read the title of this chapter. I'm also guessing that you're wondering just what rats can tell us about the value of social negotiations.

It turns out that elegant social relations can strengthen the infrastructure of brains and bodies as well as governments, news that has been shared and confirmed from rodent compatriots working in labs with clever scientists across the world. In fact, human diplomats at their clandestine meetings may want to take a note or two. For example, in adulthood, a type of play fighting seems to be used by rats to assess and manipulate social competitors and to determine if a rat should maintain a subordinate social position with its so-called partner or if it's time to go for dominance. By engaging in a bit of a social dance, consisting of gently biting the nape of the neck and rolling around or rearing up and boxing a bit with their paws, the animals gain the information they need to make their next move. What's so impressive is that no tissue damage typically results from these encounters—no bruises, no blood, no broken bones. This is not always the case with similar types of human assessment encounters.

The first time I read about the use of play fighting as a form of social assessment in adult male rats I remember sharing with my students that it was like a rat version of playing golf. Think about it. Two competitive but somewhat friendly men (or women) from the office spend the day assessing each other's information about the company's inside track. Each competitor is assessing confidence, intelligence, commitment, and athletic skill, leaving with the necessary information to determine who is dominant and who is subordinate—all under the guise of a game. The rats don't need to buy expensive golf clubs or pay green fees to do this; just a quick romp in the laboratory corncob bedding does the trick.

The sophistication of rodent social interactions extends beyond playful diplomatic meetings. They have revealed to scientists that they use tactics such as social reciprocity to ensure that they will receive assistance from their rodent friends in time of need. Further, there is a plethora of evidence suggesting that, even in rats and mice, misery loves company, and misery is relieved by the company of rodent friends. In short, medical benefits accompany living in the presence of a supportive rodent community. To be clear, these health benefits are far from subtle; if a drug could produce the effects of affiliative social contact, it would most definitely be a blockbuster. Finally, recent research confirms that mice indeed feel each other's pain, a finding that enables us to learn more about pain and healing from these intriguing animal models.

Let's start with one of my all-time favorite rat behaviors: juvenile play fighting, or free play of the rough-and-tumble variety.

A Play-by-Play Guide to Building a Healthy Brain

If you haven't seen two juvenile rats play, you've really missed out on a wonderful display of social interaction. The young rats approach each other, one climbs on the back of the other rat triggering the other rat to roll on his back, resulting in a perfect wrestling pin. The approaching rat play-bites the opponent's belly, then the pinned rat rotates so that the rat doing the pinning is now pinned himself. A chase typically ensues, tails are pulled, squeals are emitted, then the whole process begins again. Sometimes I like to think about a dramatic score playing in the background, something like Beethoven's Fifth Symphony, emphasizing the dramatic pins, rolls, and abrupt freeze responses, all followed by a continuous chase. The rats' ability to focus on each other so fully is captivating. This play choreography is so consistent across rats that there's no doubt strong biological predispositions exist for this behavior . . .

> The rats' ability to focus on each other so fully is captivating.

evolutionary precursors, perhaps, for tag, wrestling, and tackle football.

One of the industry's most creative and bold neuroscientists, Jaak Panksepp, currently at the Washington State University, has thought a lot about play behavior and the function it serves in mammalian development. When juvenile rats play in their safe cages, this behavior is not very risky. When they are in the real world, however, play expends valuable energy, distracts the animals from potential threats such as predators, and—with all the rolling around and quick turns making up the chasing sequences—presents an opportunity to get hurt. Isn't this a rather costly and risky behavior? Is it necessary for normal development?

One of Panksepp's earlier studies on this topic set out to determine the value of play by investigating the amount of makeup or compensatory play in animals maintained in social isolation. Thus, in this study, one group of young rats was housed in a social environment with ample opportunities for social play whereas the other group was housed in isolation with no opportunities for social play. When given the opportunity to have a play date with another animal, the isolate-housed, play-deprived animals played considerably more than the socially housed rats. Panksepp suggests that the motivation, or urge, to play is a natural function of the nervous system. When this behavior is limited, a type of rebound effect occurs to compensate for the previous deprivation.

Panksepp's investigations into this behavior have revealed that natural opiates seem to mediate play, perhaps contributing to a natural high as they engage in rough-and-tumble play. When the rats receive a drug that blocks the effects of endogenous opiates, play subsides; however, when given an opiate booster in the form of morphine, play is enhanced. Looking closer into the brains of the players, one study investigated the expression of the protein fos in rat brains. This is a favorite technique that I use in my lab as well because fos is produced in brain cells that were most recently active, kind of like doing a mini-PET scan on the brains of the animals to determine exactly which brain areas were active, and to what degree, in the animals of interest. Fos expression during play lights up the cerebral cortex, a brain area involved with processing information about the environment and body; in addition, other areas involved in motivation and learning are activated.

You may recall that the brain's outside cortical covering is its most highly evolved complex component. It is interesting to see this evolutionarily advanced area come alive in rats engaging in rough-and-tumble play. Panksepp has called the cortex the "playground of the mind." So it seems that play behavior may be the perfect nourishment for the rich neuronal headquarters of complex behavioral responses, especially during an animal's early years. To further support this statement, Panksepp's group has also found increased release of neurotrophic growth factors, chemicals similar to a form of brain fertilizers, resulting in the growth of brain cells in rats allowed to engage in play.

Learning that rats seem to make up for lost playtime suggests that this behavior has a functional relevance in these mammals. Play-deprived rats do not develop effective regulation of aggressive responses once they become adults. It seems that the practice of playfully testing the social waters as juveniles likely leads to more effective social communication as adults. Further, based on his research, Panksepp speculated that pro-social play fosters the development of the brain's regulatory functions, a finding that likely extends to human children: "If animal data are a valid guide, abundant play will facilitate maturation of the frontal lobe inhibitory skills that gradually come to regulate children's impulsive primary-process emotional urges."

If ever we have neglected to listen to the rodents' wisdom it is within the context of these play studies. Although still in its infancy, the research clearly points to the importance of play in the maturation of a healthy brain. Not only have we not embraced the inclusion of rough-and-tumble free play but we are systematically removing it from our children's behavioral repertoire. From 1989 to 2000, the number of elementary schools reporting at least one hour of recess decreased by nearly 30 percent. In an attempt to adhere to the mandates of the No Child Left Behind Act, time committed for recess seemed to be superfluous and was one of the first programs to be cut. And, to make matters worse, even when the programs haven't been cut, changes have been made across the country. A recent *USA Today* article titled "Not It! More Schools Ban Games at Recess" reported a trend for schools to ban games such as tag, claiming that they were too dangerous because kids were running into one another. It's not as likely that the structured indoor games

> The research clearly points to the importance of play in the maturation of a healthy brain.

are as engaging for the brain as rough-and-tumble play; at least that's the message we're receiving from the rats.

Although counterintuitive, cutting physical education and recess programs may have had a more significant negative rather than positive impact on cognitive development. If Panksepp's interpretation of his rat data is correct, we can't help but wonder about the effect of the pent-up urge to play in children forced to sit hour after hour at a desk being as still as possible. It is perfectly understandable that, like the rats that had been deprived of the opportunity to play, when any window of play opportunity is presented, the children jump on it. Of course, academic time is important. There is plenty of research suggesting that training time enhances cognitive performance— the brain just requires several types of activation, a well-balanced diet of varied experiences, to work at full capacity.

Thus this research suggests that there is a strong connection between the development of movement and cognition. More than two thousand years ago, Plato saw this without the benefits of systematic rat play studies. He wrote that "our children from their earliest years must take part in all the more lawful forms of play, for if they are not surrounded by such an atmosphere they can never grow up to be well conducted virtuous citizens."

As the free play time has diminished in our culture, an increase in the diagnosis of attention deficit hyperactivity disorder (ADHD) has emerged. More than ten million children are diagnosed with ADHD in the United States, the highest rate of ADHD in the world. Many of these children are currently taking psychostimulants, such as Ritalin, to calm their impulses and enhance focused attention. It is interesting that psychostimulants, similar in nature to cocaine and metham-phetamine, enhance learning and focused attention in chil-dren. According to the animal models, psychostimulants have a powerful play-reducing effect, making the brains of the young rats appear more mature—more like an adult brain than an adolescent brain. Thus such psychostimulant drugs may accel-erate the maturation of our children's brains.

After reading this fascinating literature about play and classroom behavior, Caitlin Blake, an undergraduate student, approached me about conducting a study to learn more about the interconnections between movement and the ability to con-centrate on a new task. Of course, we turned to the rats. Blake had two groups of young juvenile male rats; all were housed in isolation. For two thirty-minute periods each day, one group had an opportunity to play with a rodent friend whereas the other group was placed in another cage with a partner, but in this group, a screen barrier kept them from playing. They did sit very close during the non–play date, touching as much as possible through the mesh screen. The daily play dates contin-ued for two weeks, with alternating play partners just to keep it interesting.

The point of all those play dates was to determine if play had an effect on an attention task known as the attention set-shifting task. Bear with me as I describe it; it's so complicated that I've often commented that I don't think I could success-fully master such a demanding task. The rats, however, can solve it with little effort. Initially, rats were placed in a new test-ing apparatus that had two containers, each with the corncob bedding they were used to encountering in their home cage. One container had a coconut scent and the other a vanilla scent. In this phase, called simple discrimination, the rats had to learn that the coconut-smelling pot always had the coveted Froot Loops buried in it. In the complex discrimination task, we confused the rats a bit by putting shredded paper as the medium in one pot and yarn in the other. The idea of this complex discrimination task is to ignore the altered bedding media and to continue to seek the coconut scent. In the next phase, we taught the animals that they still needed to pay attention to the scent, but the scents were changed to citrus and chamomile. The rats learned that chamomile pots contained the Froot Loops. This *intra*dimensional shift required the rats to ignore a previously well-established association so that a new predictive association could be learned—difficult stuff for a rat . . . and a scientist! Finally, in the *extra*dimensional shift phase of the study, the rats had to let go of their affection for odors and focus on the type of bedding, small marbles and tiny little buttons in this case. The rats had to learn that the marble filling was the new cue for the Froot Loops. Throughout all the tests, the positions of the pots were changed so that the rats couldn't merely learn to always choose the pot on the left or right; thus position was another changing variable that had to be ignored. Are you still with me? If so, then you are likely a product of a childhood filled with boisterous play. Essentially, we were able to assess how readily rats learned relevant associations and how focused they remained on these critical associations when distracting cues were presented.

Would just two play sessions a day for two weeks give the player group an advantage in this challenging cognitive task? During the more mundane discrimination tasks, the answer was no, the nonplayers performed just as well as the players.

But things got interesting as the task got more complex. This is where the players pulled ahead, outperforming the nonplayers by getting to the Froot Loops faster, making fewer errors as they decided which pot to approach.

Since we viewed this task as a rodent version of assessing ADHD, or resilience against ADHD, we made the task more interesting in the final testing session. We placed the pots in the same apparatus but this time the walls were covered with a funky black-and-white paper. Throughout testing, the rats had to adjust to different odors, different pot fillers, different pot positions, and now distracting wallpaper. We didn't place a piece of Froot Loops in either pot this time, hoping to activate the problem-solving circuits in the rat brain. We examined the fos protein (mentioned earlier) to determine the level of activation in the hippocampus, a brain area involved in learning and memory. The results were very interesting. The player rats had significantly more neural activation as they tried to solve this unsolvable problem in the final distracting environment. Thus both behavioral and brain adjustments were improved in our rodent players, confirming the value of their two weeks of play dates. I wonder if flash cards and videos would have stimulated the brain in comparable ways.

Although we hypothesized that the play rats would perform better in this demanding cognitive task, we were surprised at the results. Physical acts made up of rough-and-tumble play facilitate performance in challenging cognitive attention tasks. The rats' ability to discern the most salient cues in the midst of competing and changing information speaks to the integrated nature of the brain and how important it is to receive a daily dose of important life experiences, especially while the brain is still developing.

As a parent, I feel an anxious pang when I think about how different my childhood experiences were compared to those of my children. My husband and I have provided more cognitive and cultural experiences for our daughters than we encountered, but I'm not sure that is better than my childhood filled with long days traipsing through the woods in my Mobile, Alabama, neighborhood—swinging on vines, climbing trees, and trying to find every single living animal possible. My days were spent playing outside with friends, but today things are different. I'm simply too frightened to let my children disappear for long periods of time with no adult supervision. I also wonder about the seduction of the many electronic devices now available. Children spend countless hours texting with their cell phones and keeping track of friends on Facebook, all without having to leave their bedrooms. These devices distract our children from actual physical encounters with their peers—forget rough-and-tumble play, they don't even hear each other's voices anymore. I have reluctantly provided my daughters with these electronic opportunities, with limitations of course, because our culture has shifted toward a love affair with minimal social contact, and I worry every day about the consequences.

In his book *Last Child in the Woods*, Richard Louv expresses similar concerns with the changing play patterns in contemporary children. In the 1980s, Louv interviewed families for a book he was working on and, for that project, he was interested in the children's perception of nature. He was struck by a comment by a fourth grader in San Diego: "I like to play indoors better, 'cause that's where all the electrical outlets are." None of us knows the actual outcome of these shifts in our children's play repertoires because the experiments investigating these drastic changes in childhood play over the past century have not been done. In the meantime, I will continue to keep my children as physically fit as they are mentally fit and try to encourage actual encounters with their friends whenever possible. Perhaps I should think more seriously about replacing gymnastics and volleyball with tag and wrestling!

Rat Reciprocity

The presence of the wonderfully choreographed play responses characterizing the complex interactions between two young rats suggests that not only are the rats aware of the presence of another animal, they are able to anticipate the responses of the other as they initiate play responses. And the same could be said for a male rat approaching a female rat before a sexual encounter or a maternal rat approaching a crying rat pup so it can suckle. Consequently, these social revelations probably don't fall in the groundbreaking category in the world of mammalian responses.

But the question to be addressed here is whether physical touch is related to the topic of rodent social awareness. In Chapter 2 I suggested that rats demonstrate *metacognition*, an awareness of what they do and do not know . . . pretty heavy stuff for a rodent. Now I'm going to rock the boat again and discuss some research suggesting that rodents have a sense of reciprocity and even exhibit signs of empathy. I'm aware that these are potentially dangerous anthropomorphic waters, and we need to be careful to not abandon the law of parsimony as we interpret these studies. That is, if the behavior can be explained by simplistic learning mechanisms, it's not appropriate to claim higher functions. So, let's proceed cautiously.

The first case of potential social reciprocity, kindness if you will, comes from the labs of Claudia Rutte and Michael Taborsky at the University of Berne in Switzerland. These researchers trained rats to pull a stick attached to a tray so that it would become flush with the bottom of the cage—all this to retrieve an oat flake. (As an aside, I was shocked to see that the Swiss rats will work for a mere oat flake . . . my rats in Virginia demand more decadent rewards such as Froot Loops!) After a few quick trials to make sure the rats knew how to operate the oat delivery system, the researchers reconfigured the system so that when the rat pulled the stick, the tray was delivered to the far side of the cage and a wire screen was added to create two chambers. Thus, when another rat was placed in the cage, the act of pulling the stick became a generous act that delivered the coveted oat to another rat.

Initially it was interesting to see that rats would indeed work to deliver food to another rat, a sister in the initial phases of the experiment. After learning to give, the rats gained experience on the receiving side. The crux of this study was to determine if the rats would be more generous after receiving oats from another rat. Human studies have suggested that we're more likely to give after being a recipient of something of value. For example, people who found a coin in a public telephone (obviously a dated study) were more likely to help someone who dropped his or her papers. You may recall leaving a larger tip when a waitress slipped in a free dessert—simple tit-for-tat kind of stuff for us—but does this social principle operate in rats?

The answer is yes. Even when paired with an unfamiliar rat,

the odds that an animal would pull the stick to deliver the oat to the other rat increased by 20 percent after that rat had been on the receiving end of the oat delivery system. There you go, rat tit for tat. The authors referred to this response as *generalized reciprocity* in the sense that even if the rat in the cage didn't deliver the oat to him in the past, just being the recipient of oat treats increased the giving act, regardless of who was receiving the treat.

We'll travel from Switzerland to the Canadian laboratory of Jeffrey Mogil for the next lesson from socially advanced rodents, this time of the mouse variety. Mogil is interested in developing animal pain models that will be informative for pain management in human patients. The Swiss study suggested that positive social interactions were contagious; Mogil was interested in the opposite, the contagion of threatening painful responses. In a study published in the prestigious journal *Science*, Mogil's team boldly used the title "Social Modulation of Pain as Evidence for Empathy in Mice." Empathy? In mice? Whoa. That's big.

To assess pain in the mice, the researchers injected a diluted acetic acid in the rodents' bellies. The injections resulted in abdominal pain, prompting the mice to exhibit writhing responses. Past research focusing on a different type of injection had shown that rats would lick their paws more than usual after an injection of a noxious agent in the same area. It is interesting that animals that received a low dose increased their licking when they were housed with a mouse that received a higher dose. In the acetic acid assessment, when two animals were tested at the same time, increased writhing responses were observed compared to when an animal was tested with an animal receiving no acetic acid injection; thus it appears that the presence of another animal in pain facilitates or enhances the perception of pain. These mice were indeed feeling each other's pain. In a clever series of studies Mogil's team found that vision was an important part of this process; when an opaque screen separated the mice, the effect was lost.

> These mice were indeed feeling each other's pain.

But what exactly were the mice seeing? This question prompted another line of research in Mogil's laboratory: They investigated the communicative value of facial expressions in rodents. Recently, when I was presenting some research about emotional responses in rodents at the annual meeting of the International Behavioral Neuroscience Society held in Sardinia, Italy, I included a slide intended to be a bit funny. As a spin-off of the popular Fox television series *Lie to Me*—a drama written to parallel the research of Paul Ekman who has conducted research tying specific facial expressions, or microexpressions as he calls them, to specific emotions—I created a PowerPoint slide of a rodent version of facial microexpressions. The six facial shots of rats were accompanied by six different emotion labels under each facial image, such as anger, happiness, or lust. The funny part was the fact that each rat face was exactly the same; the idea of discerning different emotions by looking at the fur-covered rat face seemed like science fiction to me, perfect material for an interspecies nerdy presentation joke. And it did produce some laughter. However, my colleague Kim Huhman, a trusted behavioral neuroscience friend since our graduate school days, politely suggested after the session that I read Mogil's research—that it may change my opinion about my joke slide. Oops . . . I obviously needed a dose of whisker wisdom on this topic.

Mogil's team has successfully identified mouse facial microexpressions conveying the pain response, made up of partially closed eyes, bulging nose and cheeks, and position changes of the whiskers and ears. If vision were important in the transmission of information about the experience of pain, these facial expressions could very well be the means by which the important messages are transmitted. Of course, it is also likely that these responses mitigate pain in the animal itself, and they probably do, but the existence of a two-way street, being able to express pain to a confidant as well as being able to feel another's pain is likely a valuable ability, one that began before humans ever entered the evolutionary scene.

Of course, claims of phenomena such as reciprocity and empathy existing in rodents are met by scientists with a healthy dose of skepticism, as well they should. Bold researchers have made convincing arguments against such findings being the result of more simplistic acts of simple associative learning or social modeling. Although we will remain cautious with our interpretation, the findings suggest more of a mammalian continuum with abilities often thought to be unique to humans. Whereas humans possess the most complex versions of reciprocity and empathy, components of these responses can likely be investigated with rodent models. Such research can inform our human lives. For example, if being in the presence of individuals expressing pain intensifies one's own experience of pain, then configurations of treatment and waiting rooms in medical facilities may need to be reconsidered.

Kissing Booboos and Social Buffering

As a parent, I never cease to be amazed at the power of touch in healing what ails us. Regardless of how bad a scrape or bruise is, it always seems that a gentle kiss and a compassionate and reassuring stroke through the hair have miraculous effects. The crying wails remit and the tear floodgates close. I don't know of any ointment or pill that is as effective as the practice of kissing booboos.

Does social contact really mitigate our pain? If so, how does this happen? And exactly what health benefits should we expect from meaningful and supportive social contact? These have been central questions on the minds of several researchers who have turned to rats and mice for the answers.

Several years ago one of my undergraduate students, Helen Ashley, was interested in the role of social contact in wound healing. We thought it would be interesting to use a very social mouse, *Peromyscus californicus*, a monogamous species, for this study. If social contact were important, we thought that a highly social species would be the most sensitive to the effects of social manipulations. We used males and females but didn't house them together to avoid complications associated with variable sexual contact, so sisters were housed together, brothers were housed together, and we included male and female isolate-housed groups. We exposed half the animals to the unpredictable chronic stress model described in Chapter 5 and, after about four days, we lightly anesthetized all the animals and used a dermal punch to create a small wound on the animal's back. We monitored the wounds very closely; we took pictures so the wound sizes could be quantified each day and compared to the original wound size.

For you experimental design aficionados, we were interested in the effects of three variables on wound healing in this study: gender of the mouse (male vs. female), type of housing (social vs. isolate), and stress condition (chronic unpredictable stress vs. no stress). I think it was safe to say, however, that we were most interested in the social housing variable. Even though stress typically affects immune function, it didn't have an impact in this study. The housing variable, however, did prove to be interesting because social housing facilitated wound recovery on several of the observation days. What about gender? On the last day of observation, the females' wounds were virtually completely healed whereas the males' wounds were still apparent. So, in this study, above and beyond the presence of chronic stress, the most influential variable determining the rate of wound healing was having affiliative social support, especially if you're a female mouse.

Another student, Erica Glasper, conducted a similar type of study during her graduate program at the Ohio State University working with behavioral neuroscientist Courtney DeVries. In addition to investigating wound healing in the monogamous California deer mouse (as we did), they also looked at a nonmonogamous species, the common deer mouse, a mouse species known for its tendency to play the field. DeVries and Glasper conducted several interesting experiments, reporting that the monogamous species experienced faster recovery rates than the nonmonogamous species when housed with a partner. This benefit seemed to demand actual physical contact, however; when the animals were separated by a double-screen barrier, the immune benefits ceased. I suppose it's difficult to maintain that therapeutic cuddling through a screen. Thus these data suggested that, above and beyond visual, auditory, and olfactory signals released from animals in social housing, sustained contact seemed to be the most critical for the healing benefit in these monogamous animals.

DeVries has written extensively about potential mechanisms underlying the social buffering effects observed in her laboratory rodents. In addition to facilitating wound healing, she has also observed facilitated recovery from stroke. Although there are many species and individual differences, it looks like positive social interactions suppress stress hormone levels. Stress hormones are important for defending our bodies against a plethora of threats but, as discussed in Chapter 5, prolonged periods of elevated stress hormone levels often come with a cost, a cost typically paid by immune functions. The neuropeptide oxytocin appears to play a critical role in pair bonding in several rodent species and may diminish the toxic effects of chronic stress. In the monogamous prairie vole, for example, circulating stress hormone levels are up to ten times higher than reported in rats—and finding that special someone (establishing a pair bond) decreases stress hormone levels to half their baseline levels within a mere fifteen minutes. In this case, love at first sight is sustained by increased oxytocin activation followed by plummeting stress hormone levels, an effect that likely has rewarding, calming effects.

A group of rats in the laboratory of Yasushi Kiyokawa and his colleagues at the University of Tokyo provide further support of social buffering during stressful situations. Working with an animal model of fear, these researchers used classical conditioning, à la Pavlov's bell and salivating dogs, to induce stress in rats by pairing a sound with mild foot shocks. The intensity of the stress response to the sound alone once it had been associated with the foot shock was measured by behavioral observations (rats tend to freeze when they are very frightened) in addition to physiological measures (such as body temperature changes) and neural activation in brain areas associated with fear. The scientists tested the effects of having a roommate before the fear conditioning test as well as the effect of having a buddy with them when the conditioned fear sound was presented. Each social condition had a mitigating effect on the stress responses, but what was most interesting was that the combination of the two social factors—having the social support in the form of a roommate before the challenge and experiencing the stress with a familiar rat—totally wiped away any evidence of the conditioned fear tone being stressful. This finding suggests that, in some cases, positive social contact goes beyond buffering the negative effects of physical and psychological threats. In this case, social support completely erased the fear footprint on the body and brain.

> In this case, social support completely erased the fear footprint on the body and brain.

As the threats become more perilous, social contact continues to offer impressive protection. Martha McClintock's laboratory at the University of Chicago confirmed that social support is not a passive endeavor. Using an elaborate behavioral scale, these researchers found that rats engaging in reciprocal affiliation (initiating social interactions as often as receiving social prompts) during a mild stressor exhibited a lower stress hormone peak. The researchers also investigated the development of mammary tumors in their rats. The multiple breasts and nipples in female rats make this species an excellent model for investigations of the development of mammary tumors. Similar to humans, rats spontaneously develop both malignant and benign tumors. The rats that were the most affiliative, initiating more social interactions, developed tumors later in their lives than the nonreciprocating loner rodents. And, even more impressive, the high-reciprocity rats lived 30 percent longer than the low-reciprocity rats.

In another study, McClintock's team confirmed the important relationship between social support and the development of spontaneous mammary tumors by examining tumor development in single housed and social housed rats. By middle age, which is about fifteen months of age if you're a rat, about 74 percent of the rats had developed at least one palpable mass with no difference between the two groups. Social status did, however, have an effect on the number of tumor masses, with a 135 percent increase in the socially isolated animals. Concerning the nature of these mammary masses, isolate-housed rats had a 3.3-fold increased risk of developing breast cancer, and 50 percent of the isolated rats developed some form of carcinoma compared to 15 percent of the group-housed females.

In closing, the information discussed in this chapter confirms that rats are very social creatures and their sophisticated social interactions preserve their well-being just as social diplomacy preserves the security of nations. The well-orchestrated rough-and-tumble play helps build the brain, another form of central intelligence. Being able to gather the most relevant information about social contacts throughout a rat's life is critical for survival. Further, just as rats look to each other for pointers on which foods to consume or pass by, they just might plan for the future by reciprocating a kind response, especially in the climate of previously receiving kind gestures from other rats. These social interactions have very real health effects in the rats, measured by solid medical endpoints, such as lower stress hormone levels, longer life spans, and fewer malignant tumors.

Thus, when asked, the rats answer our questions about the value of positive social interactions with clear answers, all pointing to therapeutic advantages that rival our best drugs and surgical innovations. This is no new news for humans. But the rats have provided a means to investigate the exact mechanisms and specific endpoints of socially mediated health benefits, especially as our culture is rapidly changing to accept less and less face-to-face, skin-to-skin social interactions. Perhaps our rodent friends can provide a valuable dose of wisdom for us humans: Don't lose sight of how important it is to stay in touch with one another; life is better when we do, in many, many ways. If simple screen barriers block the powerful benefits of social contact in rodents, we should consider the isolating impact our video screens have on our social interactions. This social-deprivation experiment is ongoing in our society, and the impact will not be fully known or understood for several decades to come.

CHAPTER 7

||||||||||||||||||||||||||||||||||||||||||||||||||||||||||||||||||||||||

The Perils of Bad Hair Days

In retrospect I probably overreacted. After reading all the scientific literature about the value of good old-fashioned rough-and-tumble play and the potentially threatening nature-deficit disorder, my daughters' requests for the latest and greatest electronic equipment for Christmas were not met with an approving nod. How could I reinforce the demise of my daughters' emotional resilience? I recalled the interest my oldest daughter had in a chinchilla we had seen and . . . well, the rest is history.

After researching chinchillas, arguably the most glamorous of the rodents, I thought it would be good for the girls to have something natural rather than technical to care for. I found a reputable breeder, ordered two female chinchillas for Christmas Eve, somehow managed to persuade my husband that having these rodents for the girls to care for would be a valuable, even brain-enriching experience, and delivered the goods to my confused daughters on Christmas morning. Instead of an iPhone, they got chinchillas—wonderful little rodents that required attention, cleaning, hay, the whole works. It just felt better than an electronic device. After learning that these animals live up to twenty years, my husband felt a little bamboozled, but I adore little Sadie and Sophie and continue to learn from their extraordinary behavior, especially their fascinating hair care.

If you've ever had an opportunity to touch a chinchilla, you know they have unimaginably soft hair that's very thick: about sixty hairs per follicle (humans have one hair per follicle!). Of course, this trait led to the demise of many wild chinchillas as they were killed for the fur trade. With this hair comes the responsibility of proper care. It's too dense to clean efficiently with water so the preferred grooming method for chinchillas is a dust bath. These South American animals roll around in volcanic ash in their native habitats to protect their fur. Our Sadie and Sophie have a bathhouse filled with very fine sand that we place in the cage each day. This behavioral response gets me every time. The chins run to the bathhouse as if it were stocked with cocaine, push each other out of the way, plop on their backs with their short little forearms and awkwardly large feet, and roll furiously—all to get a good dousing of the fine sand on their fur. It's the perfect rodent desert spa!

I admit that I'm more enamored by these behaviors than anyone else in the family. I am amazed each day to see how motivated these rodents are to maintain appropriate hair care. Possessing a finely coifed fur coat is certainly a high priority response for chinchillas. Why is the condition of their fur so important? The dust keeps the fur free of oil and dirt, essential for maintaining body temperature and providing a barrier against parasites.

You may be thinking that the chinchillas' good hair hygiene makes perfect sense, but what about rats? Disgusting, filthy rats? Surely they're not as coif-conscious as the chins. Think again. Rats spend up to 50 percent of their waking time grooming their fur. How much time have *you* spent bathing and combing your hair today? If rodents invest so much time tending to their fur, you may be wondering what they have that we don't; what are we missing? As I observed in the chinchillas, there is biological relevance for striving for good hair days. In addition to providing good hygiene, rat grooming is important for temperature regulation, the dispersal of key chemicals from the skin that send specific messages to other animals about health and reproductive status, and, interestingly, grooming can be used as a coping strategy to diminish stress.

> There is biological relevance for striving for good hair days.

We can learn some valuable lessons from rodent grooming habits. In this chapter, we'll see how complex and involved the grooming response is in the rat—a lot of brain power is devoted to this response—and how this information may help scientists discover cures for movement disorders such as Parkinson's disease. We'll also learn from some Canadian rats about the value of grooming the kids and how this information can lead to a heightened resilience against the emergence of anxiety disorders down the road. Of course, there are limits to most good things. Certain strains of mice have taken the adaptive response of grooming a bit too far by becoming grooming bullies and barbering their cage mates. Yes, Mohawks, interesting bald patterns, and other unique hairstyles are being observed in laboratory cages across the world. They range from trendsetting to upsetting. After we work our way through some rodent hair tales you'll have a new appreciation for those proverbial good hair days.

Rat Grooming 101

Let's review the typical grooming session of a rat. You may want to grab a notebook, because it's complicated. The typical five-second grooming sequence contains about twenty-five movements that can be divided into four phases. The grooming sequence is *fixed*—that is, it proceeds in a very predictable pattern. On a general level, the grooming sequence, including wiping, licking, and scratching, travels in a cephalocaudal direction, meaning that it moves from the face and head (cephalo) toward the tail (caudal). During Phase 1, the rat uses its paws to make up to nine circular strokes over its nose and vibrissae (whiskers) for about a one-second duration. Phase 2, lasting only a quarter of a second, comprises small asymmetrical circular strokes that get larger with each stroke as the paws start approaching the area of the ears. Phase 3 consists of very large circular motions, this time running the forepaws at a distinctly slower pace across the backs of the head and ears. Finally, in Phase 4 the rat adjusts its posture so it can lick its back and sides, working its way down to the tail. In addition to these fixed sequences, rats will also exhibit incomplete, variable grooming sequences. According to reputable rodent grooming researchers, such as Kent Berridge at the University of Michigan, rats will engage in up to fifteen fixed chains of grooming behavior per hour (lasting up to seventy-five seconds). The variable chains, however, occur more often, with the total duration of these bouts being about twenty times greater than observed in the fixed chains.

This wealth of grooming knowledge hasn't come easy to researchers. It takes labor-intensive moment-by-moment behavioral transcriptions of rodent grooming videotapes to generate these data. It is interesting that researchers speak of grooming sequences having syntax—a predictable pattern of sorts as observed in language. Once the grooming sequence begins, each behavior in the sequence can be predicted with an accuracy rate exceeding 90 percent. And the basic component of this sequence has been conserved across several species in addition to rats, including hamsters, mice, guinea pigs, and squirrels. Although many brain areas are involved, the primary areas regulating the predictable movement sequence are the basal ganglia, the central part of the brain involved in movement regulation, and the brainstem, the brain structure that regulates basic functions such as respiration and muscle tone.

You may be thinking that a few tidbits about rat grooming may be mildly interesting, but why all the hoopla over rat hygiene? With seventy thousand hair salons in the United States and a combined nineteen billion dollars in hairstyling sales, a detailed analysis of this behavior may be valuable for understanding that profitable industry. But what can we learn from

What can we learn from researching rat hairstylists?

researching rat hairstylists? It turns out that we can learn plenty about the brain and the behavior that emerges from the neuronal networks. Put differently, learning about the mechanisms of this robust response in rodents may reveal interesting insights about our own behavior. Grab a brush and keep reading!

Now that the behavioral sequence and corresponding brain areas have been mapped out, scientists can learn more about how the brain initiates and maintains motor sequences. Although we may not exhibit a stereotypical four-phase grooming sequence in our morning shower, we have plenty of stereotyped motor sequences that get us through the day. Walking, talking, scratching, typing, knitting, even sneezing and yawning, all require well-established motor programs. Researchers have found that when the basal ganglia and closely related areas, even small portions of this area, are damaged in rats, the number of grooming chains that the animals actually complete drops off by more than 50 percent. The rat has no problem attempting the sequences but just can't follow through.

This scenario is reminiscent of a human disorder with similar symptoms, Parkinson's disease. Patients with this neurological challenge seem to have all the neuroanatomy to decide and position oneself to trigger a particular behavior, but something goes wrong in the implementation and execution. In Parkinson's disease, a brain area known as the substantia nigra is compromised. The consequential drastic reduction in a critical neurochemical, dopamine, is the culprit for the movement false starts. This neurochemical is important for movement as well as other functions, such as brain reward (as discussed in Chapter 3). Sure enough, reducing the available dopamine in the rats compromises their grooming behavior.

The study of rat grooming also helps us better understand other diseases of the brain's motor systems, such as Huntington's disease and Tourette's syndrome. The rodents' bad hair days may represent significant brain disruptions and should be taken very seriously. With the help of rats, we hope to learn more about rebuilding the neurobiological system to provide new options for the millions of people suffering from these disorders.

Taking a Licking Leads to Prolonged Ticking: The Importance of a Mom's Hair-Care Strategies

One of the most celebrated lines of rat research in behavioral neuroscience these days involves the work of Canadian neuroscientist Michael Meaney and his obsession with the number of times a mama rat licks her young. Sounds pretty trivial, doesn't it? On the contrary, you have no idea how much we have learned about some fundamental philosophical issues such as the age-old nature–nurture question by studying the effects of a mama rat's grooming habits. Let's start this story with a little history.

Seymour "Gig" Levine, who spent most of his career at Stanford University, conducted seminal research half a century ago exploring the effects of maternal separation on the developing stress response in young rats. Although being separated from one's mom seems stressful and, at first thought, may seem harmful to the developing rat, Levine learned that this experience was beneficial in the sense that it diminished the behavioral and hormonal responses to stress later in the rat's life. Perhaps, the rodent moms were teaching us a bit about tough love, distributed in small, transient doses. As adults, these mildly stressed rat kids were protected from overreacting to stress when it wasn't necessary, a characteristic leading to less toxic allostatic load, discussed in Chapter 5.

As Levine attempted to understand the counterintuitive finding of beneficial effects of maternal separation, it became apparent that this brief separation altered the rat mom's response toward the offspring. Upon being reintroduced to the cage, the mom spent more time grooming the pup. It was as if she were thinking, "Where have you been? Come here and let's get you cleaned up."

I suppose it's time to inform you of a little rodent personal hygiene secret: Rat pups can't go to the bathroom on their own—they require a little licking in the anal–genital area to stimulate that response . . . definitely something only a mom could do. Thank goodness for some distinct species differences in this behavioral category, and the creation of baby wipes! Anyway, it became apparent that the increased maternal attention seemed to be shaping the offspring's stress response. It turns out that the mom's licking is a major source of stimulation for the rat pups and that the tactile stimulation regulates the developing hormone and cardiovascular systems. At the time this research was first being published, defense responses (for example, the stress response) were thought to be fixed or carved in stone, definitely heavy on the nature side of the nature–nurture continuum. Could it be that moms could shape the stress response, a response so essential for mental health, by the amount of licking and grooming directed toward the pup during the first couple of weeks of life?

Most of the data contributing to this rat tale have been subsequently conducted by Meaney and his McGill University team. Eventually it became apparent that the researchers could skip the maternal separation part of the investigation and simply look at natural variations in maternal licking and grooming. Sure enough there were natural differences; some rat moms (perhaps rat versions of human soccer moms) were categorized as high-licking and -grooming moms, whereas others fell in the category of low-licking and -grooming moms. Thinking back to my awkward-looking elementary school class pictures, I am quite certain that such grooming differences exist in human moms as well. Most of the children's hair is unremarkable, combed straight but neatly arranged, but then there are the extremes: the kid with the hair going in all directions who looked like he just tumbled out of bed and the kid with the well-prepared coif of perfectly adjusted curls and bows and ribbons. Lots of individual differences for sure.

To facilitate this research, Meaney's team started breeding high- and low-licking strains of rats. Sure enough, when rats raised by high-licking moms were tested later in life, they had a healthier stress response than rats raised by low-licking moms. This led to healthier brain areas, especially the hippocampus, involved in learning and memory. Equally fascinating: When rats born to low-licking moms were adopted and raised by high-licking moms, they also experienced the beneficial transformation of their stress responses. Now it looks like the stress response is landing more on the nurture side of the nature–nurture continuum.

Decades of research on this topic have produced results that are nothing short of astounding. Not to put too much pressure on hardworking mothers out there, but this research provides strong evidence that the mother's nurturing abilities determine the eventual stress response phenotype, or final profile characterization. Tested as adults, the offspring of high-licking mothers have decreased levels of the precursor hormone triggering the stress response (known as corticotrophin-releasing hormone, or CRH) and lower plasma levels of stress hormones such as glucocorticoids. Several biological markers also indicate a more efficient response to the stress hormones in the hippocampal brain area, modifications that lead to a more sensitive feedback system. These rats also exhibit less severe startle responses and are more exploratory in novel environments—we've already discussed the advantages of being a bold rather than a shy rat. When placed in a threatening situation, the offspring of low lickers exhibit more defensive behaviors, such as excessively burying the offensive stimulus, a behavior that could easily develop into anxiety-related compulsions. The list goes on, but you get the point. If given a choice, choose a high-licking mom! And if you are somewhat mentally healthy, remember this when you are picking out a Mother's Day gift next year.

Meaney's efforts have recently been directed toward an explanation of these results. How does licking translate into a modified stress response in the offspring? In a word, through *epigenetics*—taken from the Greek word *epi*, meaning "upon," and *genetics*. We now know that behavioral events can alter certain aspects of the genetic code, or the nucleotide sequences, without actually reorganizing the DNA bases. Kind of like editing the words without changing the basic structure of the story or playing a tune at a different pitch or tempo. One way this happens involves a rather sophisticated process called methylation, which amounts to having little courier chemicals known as DNA methyltransferase transfer methyl groups to the DNA. This methyl musical chairs can result in a silencing of certain genes. At one time scientists thought such genetic tweaking could occur only during embryonic development, but now we know it exists in mature cells as well. The quality of a maternal rat's licking can silence certain genes related to the stress hormone receptors in the brain, ultimately influencing the animals' stress response.

Although the specific genetic events accompanying the high- and low-liking moms' enduring effects on their pups remain controversial in the broader scientific community, these effects have been viewed as revolutionary by most behavioral neuroscientists. Different environmental conditions can silence genetic expression, which ultimately influence long-term physiological responses. Is this nature or nurture? As we're learning more about the intricate relationships between environmental conditions and genetic programming, the boundaries between nature and nurture are becoming less apparent. Both are essential. Rather than being fixed, in most cases nucleotide bases are modifiable so that environmental conditions and social interactions can shape the genome to respond most adaptively to the existing conditions. The context of our lives shapes our genome, and in the case of rats, a mother's touch shapes her pups' biological destinies. I don't remember learning that lesson in my baby classes!

The final genetics lessons rats have taught us in this area of development involve the transmissions of certain traits from one generation to the next. As we all know, that happens through genetic transmission, or does it? By using a clever research design, Meaney and his colleagues looked at the grooming habits of rats that had been raised by either a high- or a low-licking mom. Sure enough, the mothers doing the grooming, not those that contributed the genes, are the most influential in determining the offsprings' subsequent pup licking and stress responses. In an exciting article published in the prestigious journal *Science*, these multigenerational rat families made history in an article titled "Nongenomic Transmission Across Generations in Maternal Behavior and Stress Responses in the Rat."

Although this research celebrates the importance of a mom's touch, before moving on to the next section, it is important to mention some research indicating the importance of paternal investment—that is, when you can find it. For the few monogamous and paternal rodent species, evidence is trickling in suggesting that fathers make a difference as well. German neuroscientist Katharina Braun investigated the effect of removing the paternal contributions in the degu, a rodent species that resembles chinchillas. Like the California mice we work with in my lab, the male degu contributes substantially to raising the kids. In Chapter 9, we'll explore these uniquely paternal rodent species and the impact they have on their offspring.

Driven to Pulling Your Hair Out: Mental Illness and Hair-Care Practices

Remember those bold and shy rats we discussed in Chapter 5 and the picture of those same-age brothers depicted in a conference presentation that stopped me in my tracks? The shy one looked so old, whereas the bold one looked so young and perky. I didn't mention the exact characteristic in the picture that made the shy rat look so old. I was saving that juicy detail for this chapter because it was most definitely the hair. Through carrying out the complex grooming sequence discussed earlier, the rat keeps its fur quite clean, even around the top of the head and shoulders and back—those hard-to-reach places without long scrubbing brushes that humans use. To put it bluntly, the shy rat's hair was all messed up, disheveled, and an unattractive dirty yellowish beige color. The bold rat still sported shiny white perfectly styled fur. In fact, fur quality is so reliably a measure of personal neglect that it is often used as a measure of depression in rat studies. Thinking of Jack Nicholson's hairstyles in certain scenes in *The Shining* and *One Flew Over the Cuckoo's Nest* suggests that screenplay directors may have also noticed this trend. Back to the rats: Given the importance of the grooming response, if a rat stops grooming, you don't need blood work or MRI scans to know that something is seriously wrong. When human patients are depressed, less energy is devoted to personal care as well.

When French neuroscientist Catherine Belzung exposed mice to the unpredicatable chronic mild stress model of depression described earlier, the unpredictable annoying stimuli she used included damp sawdust, social stress, predator sounds, and erratic light-dark schedules. One variable she was interested in was the state of the fur. This variable was a gauge for determining the negative effect the stress was having on the rodents, mice in this case. In fact, each day the experimenters would visit the mice for a bit of a quality fur control check, looking at various body parts for fur status. The mice received a score of 0 if their fur looked healthy, and a score of 1 was reserved for bad hair days. It is not surprising that the coat state was reported to be degraded in the stressed animals. One manipulation in this study was to assess the efficacy of two types of antidepressant-like drugs (tramadol and desipramine) on fur degradation; for each of the two drugs, the fur neglect was reversed. Belzung's team further challenged the grooming abilities of the mice by exposing them to a splash test, which consisted of squirting a small amount of saline solution on their backs and recording how quickly they started grooming as well as the duration of the grooming bout. Again, stressed animals were less vigilant about cleaning their fur in this test, and again, the antidepressant-like drugs reversed the effect.

Researchers are also beginning to find that there is more to hair care than meets the eye—what meets the eye in real time, at least. When Tulane neuroscientist Allan Kalueff and his colleagues investigated the effects of being a social loser rather than a winner in life, they initially found no differences in overall durations of grooming. They assessed dominant and submissive mice in social encounters; as might be expected, the animals doing most of the chasing, biting, and pinning were deemed the dominant winner mice, and the animals doing more of the fleeing and engaging in awkward positions against the side of the wall (in which they leaned on their sides and threw their paws up in a defensive posture) were deemed the subordinant losers. After a couple of weeks of being exposed to dominant mice, a first glance at grooming responses revealed no real qualitative differences between dominants and submissives. But using special recording software and a lot of patience, researchers learned that the chronic social stress of being bullied led to more disrupted grooming chains. In other words, the sequencing of beautifully choreographed grooming response was disrupted, suggesting less focus and more disruptions in the stressed animals.

Recently, we observed a similar effect in our paternal California mice. Working with Marshall University neuroscientist Massimo Bardi, we assessed grooming responses in mice with varying paternal experience—mice who were dads, foster dads, or who had absolutely no paternal experience. When exposed to the mild stressor of an open field with a novel object, there were no differences in overall grooming latency or duration scores. However, focusing on the microstructure of the grooming response we found that, as the parental experience declined, the number of interrupted grooming sequences increased. As we'll discuss in Chapter 9, being a parent—in this case a dad—changes the parent's brain and behavior in fascinating ways.

Thus, according to the rodents—rats and mice alike— maintaining healthy hair is a sign of mental health. As with everything else, however, too much of a good thing can be bad news. Humans seem to go to extremes in their hair care—plucking or pulling their hair to excess, adopting grooming patterns that are outlandish or excessive (such as Mohawks and shaving

Maintaining healthy hair is a sign of mental health.

sport team emblems on the back of their heads). Again, these hairstyling extremes are not unique to humans; the rats have their own version of hairstyles gone bad.

One of the most intriguing rodent behaviors is the aptly named *Dalila effect*. Based on the name alone, there's no doubt that you know where this is going. When laboratory rodents are group housed, lab caretakers will every once in a while observe something odd. It happened in my lab just a few months before I wrote this chapter. The students informed me that it looked as if one of the rats had its forearms shaved. That seemed odd to the students as we don't issue any shaving tools to our rats and mice. Other reports in the literature reveal more extravagant grooming patterns, with mice emerging with symmetrically shaved ellipses around their eyes or half their face shaved or an interesting racing stripe shaved down the back. When the mice are serious, though, they go for the most valuable hair specimen, the whiskers. These whiskers are so important for rodents that they have specialized neural cell networks, called barrels, devoted to regulating each whisker. As rodents navigate their dark worlds each evening, their whiskers are actively whisking, taking in important information about structural barriers to guide their movements. Having even a few of these whiskers yanked out could have devastating survival implications for the rodents.

Exactly how does the barber mouse retrieve a whisker? And what does he do with the whisker after the surgical excision? Rodent researcher aficionado Ian Whishaw and his colleagues at the University of Lethbridge in Alberta, Canada, have used sophisticated video analysis to discern exactly how mice carry out this barbering task. Whishaw's team has published vivid pictures of barbering mice caught in the act. The barber typically positions himself on top of the customer, holding the mouse down with one paw while grabbing the whisker with its incisors. After pulling the whisker out with its long teeth, the barber mouse eats the whisker. Talk about recycling.

But there are some intriguing aspects to the Dalila effect. Not unlike the interesting and confusing aspects related to human sadomasochistic interactions, Whishaw reports that, while the barber mouse is eating the whisker, he no longer restrains the victim mouse with his paws. Yet, the mouse continues to sit there. They were surprised to observe that in all the barbering episodes, the barber was just as likely to be approached by the barbering recipient mouse as he was to approach the recipient mouse. Thus, even though researchers have suggested that this is certainly a dominance scenario, with the barber mouse, always the larger of the pair, assuming the dominant position, there seemed to be a bit of cooperation going on.

Perhaps the barbering relationships are less about cooperation and more about the recipient animal's insight that if it cooperates, it will suffer fewer painful consequences. Even so, when the barber and recipient mice are separated by a wire screen, barbering still continues. Also interesting, other researchers have observed robust barbering during sexual behavior in some mouse strains.

This barbering effect may be viewed as a form of anxiety-related compulsive behavior or a way to gain control in a situation in which an animal has little control over its environmental conditions (such as being housed in a small cage). The Dalila effect has not been observed in rodents living in natural habitats; it is specific to the confines of the laboratories. We don't live in cages, but the disconnect between the natural environments and active lifestyles of our ancestors and our contemporary lifestyles may also swing the pendulum in human responses toward compulsions in an attempt to gain even the smallest semblance of control in our hectic lives. As discussed in Chapter 3, action is a powerful antidote to stress and anxiety; however, sometimes such actions lead to additional problems. Of course, like everything else, actions should be dispensed in moderation.

So the rats have taught us another valuable lesson. The most mundane routine behaviors—such as grooming—are the neurobiological staples of life. Rodents remind us to embrace the complexity and wonder of what we see and do every day of our lives.

CHAPTER 8

||||||||||||||||||||||||||||||||||||||||||||||||||||||||||||||||||||||||||||

The Final Rose Ceremony

If you have no idea what this title refers to, congratulations! That means you have managed to escape perhaps one of the silliest reality shows crowding today's entertainment airwaves. For those of you who are familiar with this phrase, you know that this is a ritualistic ceremony that has become an iconic part of the reality TV series *The Bachelor* and *The Bachelorette*, in which a desirable bachelor or bachelorette is presented with twenty-five potential mates who are systematically weeded out over a series of artificial encounters and interactions staged for an audience of millions. The most desirable potential mates are given a rose each week, a gesture indicating their special unique value as they stand among a crowd of other potentially valuable spousal candidates. All of this drama leads to the final show and the "final rose ceremony," during which the rose is given to the most desirable mate, often accompanied with a marriage proposal, although the entire relationship has been staged for a television audience. To no one's surprise, out of twenty shows and final rose ceremonies, only one marriage (a 5 percent success rate) actually emerged from this confusing, uninformed, maladaptive mating game.

This show is saturated with cultural icons representing romance and desire: physically beautiful people, bikinis, hot tubs, gourmet meals, adventure dates, champagne. Beyond the glitz and glam, however, rats inform us of the essential elements for romance, which are definitely missing from this reality show version of a relationship train wreck. The truth is that when it comes to sex and understanding its mechanisms and functions, we know a lot more about rats than we do about humans. Because the romantics out there would rather take advice from poets, advice columnists, and fictional dramatic plots, humans are more confused than ever about relationships. The information revealed by rats, mice, and the monogamous prairie vole illuminate some of the dark mysteries of romance, reminding us of the importance of a few basic ingredients that are critical for successful relationships, regardless of your species.

The Joys of Rat Sex

I still have vivid memories of my first rodent sexual encounter. I was a rookie graduate student in the biopsychology program at the University of Georgia. Being only twenty years old, I felt inadequate as I looked around at the older, mostly male graduate students. They were already wearing suspenders and smoking pipes! I put on my most mature face and attempted to mask my southern accent in an attempt to fit in with this academic crowd. To introduce me to the rodents, my professors sent me to the rodent laboratory on College Station Road to observe some action. My undergraduate training didn't include actual research with animals, and I was excited to begin my research career.

It was clear that this experience was unlike anything I had ever encountered or even thought about encountering in my conservative southern upbringing. I found myself standing in a small, dark room with three older male graduate students in a double-wide trailer retrofitted as a rodent laboratory. We surrounded a circular Plexiglas arena that was illuminated with red lights (not detected by the rodents). The bottom of the arena was equipped with mirrors so that every move of the hamster couple could be observed and recorded. In this session, the seasoned students were introducing me to the behavior, so commentary was allowed as we watched the sex sequence.

I wasn't sure what I was seeing. Two little hamsters darting around, sniffing, mounting, grooming. The female seemed so rigid, so passive. Did she not like this guy? Why did the male keep repeating this behavior?

I obviously had a lot to learn about rodent sex.

Was he confused? Was this a learned behavior and he just needed more practice? I kept all these questions to myself and nodded when the male graduate students excitedly conveyed that they had seen the ejaculation. I had no idea how to determine when the hamster was ejaculating. I obviously had a lot to learn about rodent sex.

Fast-forward a quarter of a century and I can confidently say that I have indeed learned a lot about rodent sex through the years. During that time, my focus has shifted from hamsters to rats. Of course, the scientific literature has helped clarify what we see in the laboratory. Rats have been studied intensely, allowing us to use them as a model to discern the influence of hormones, drugs, age, and a host of other variables on sexual behavior. Based on decades of research, we know that rats follow a rather predictable sequence of events in the bedroom. It goes something like this . . .

The male rat greets the female rat by sniffing her, starting perhaps with a friendly nose sniff but quickly working his way to the anal–genital area. If the male seems distracted, the female may get his attention by darting around the cage and engaging in the ultimate rat flirtation response: ear wiggling. I still feel like I'm watching a cartoon when I see this behavior; the ears look like little helicopter propellers as the female entices the male to approach her. There is also a lot of sniffing. The female has a very short estrous cycle, one that repeats itself every four to five days. During this cycle, there is a relatively brief time, only twelve to twenty hours, that the female is sexually receptive, or in a behavioral state known as estrus. It is only during this time, after ovulation, that the female will mate with a male. In fact, if a male approaches when she isn't in estrus, she's likely to give him a swift kick. This time restriction for sexual encounters is very efficient in that it is the only time the copulatory response will lead to pregnancy. Not having access to text messages or Twitter, the males rely on their olfactory system to determine if the female is in the mood for love.

If the female is in estrus and has attracted the male's

attention, only then will the copulatory sequence begin. As the male mounts the female's back and grasps her sides, she reciprocates by assuming a posture known as lordosis, a behavior characterized by an arched back and lifted tail. During the initial mounts, the male exhibits a series of shallow, rapid pelvic thrusts. If penetration is not achieved, he dismounts by pushing the female away rather slowly. If penetration is achieved, the male makes a deeper thrust, just one this time, and quickly pushes away from the female with a lot more zest than the non-penetration dismount. This mount with accompanying penetration, lasting all of two hundred milliseconds, is known as an intromission. After an intromission, the male engages in a rather unique response—genital grooming, yes he licks his penis—in a very excited fashion. Then he mounts the female again, either with a simple mount or another intromission. In an extended version of foreplay, this sequence will continue until the male has achieved from eight to twelve intromissions. At this time, the male's behavior changes. After the next mount and penetration, the male thrusts deeply for about five times, followed by an orgasmic spasm of the hindquarters during the deepest penetration. This behavioral sequence occurs when the male ejaculates. (That's what my graduate school buddies were seeing in that dark room so many years ago.)

If an observer misses the hip flutter accompanying the ejaculation, the dismount is another sign that he has ejaculated. Instead of robotically dismounting, the male tightly clutches the female, dismounting very slowly, almost as if he were melting away. To enhance the romantic ambiance, the male likes to sing throughout this process—he has a special ultrasonic vocalization announcing each phase of his sexual performance.

There is a mating call as he approaches the female, and he actually breaks out in a preejaculatory song to announce the imminent event just before its occurrence. After ejaculation, the male emits a postejaculatory call, not to seduce the female but rather to keep her away while he rests. The entire sequence, complete with the sharp moves, distinct songs, and passionate chases and grasps, reminds me of a heated rodent tango.

Stay with me, we're not done yet. After this intense sequence the male will take about a five-minute break and, yes, begins the entire sequence again. If left undisturbed, the male will repeat this sequence five or more times. (I enjoy watching the sometimes sleepy male students in my animal behavior lectures suddenly wake up from their slumber as they hear about these rat sex studs.) As the males continue this marathon date, however, the time between the ejaculation and the next intromission, known as the postejaculatory interval, gets longer and longer as if the male were getting tired over the course of the date night. At least that appears to be what the male would like to convey; however, clever scientists know that, in some ways, this seemingly pooped-out male is holding back.

If a new female is placed in the cage, even when the male is displaying his "I'm exhausted from all this ejaculating" act, the rules suddenly change. With a new female in the area, an entirely new ejaculatory sequence begins, providing evidence that, for the rats, a little variety tends to spice things up for the male. This phenomenon is a favorite of biopsychology and animal behavior textbook authors because, according to legend, it has a rather unique namesake. According to the story, President Calvin Coolidge was touring a chicken farm with his wife, Grace, during the 1924 reelection campaign. For some unknown reason the president and his wife took separate tours. When Mrs. Coolidge noticed the sexual vigor of a prize rooster, she asked the tour guide about the number of sexual encounters the rooster averaged each day. After learning that the number was rather high—up to twenty encounters a day—she was visibly impressed and asked the guide to mention this interesting bit of information to the president when he made his tour through the rooster's pen. When President Coolidge heard this statistic, he asked if the encounters were with the same hen every time. The guide commented that, to the contrary, it was with a different hen each time. The president then asked the guide to convey that bit of information to Mrs. Coolidge.

In the real world outside of the laboratory, rats often engage in group mating. With groups of animals, the females have to effectively gain the attention of the male and the male has to focus on just one animal until he completes the copulatory series. Even so, we can learn a lot about the complex copulatory response by systematically investigating sexual behavior in the controlled confines of the laboratory. For example, when tested in the appropriate conditions, we now know that females actually pace the timing of the males' mounts and intromissions. Even though the males prefer fast-paced intromissions and ejaculations, females require longer intervals to maximize genital stimulation to ensure fertilization and pregnancy. When males are tethered in a cage and the copulation depends on the advances of the female, her optimal pacing schedule becomes apparent.

We can also use the rat sex model to learn about the potential disruptions or enhancements of certain drugs or other conditions on sexual responses. In a recent study, for example, a group of Italian neuroscientists wondered about the effects of rave parties on sexual behavior in humans. The researchers thus administered MDMA (Ecstasy) to rats and played very loud music for a specified period of time. My guess is that the scientists found it difficult to receive institutional approval to conduct this experiment on animals, yet adolescents and young adults are crowding tightly into clubs across the world engaging in these responses, except with higher doses of drugs, louder music, and warmer temperatures. MDMA suppressed the sexual vigor of the male rats and, surprisingly, the loud music mitigated the negative effects of the drug somewhat, but not to baseline levels. So the data from the rat raves suggest that such all-night parties have a negative impact on sexual responses.

All of this rat sex reporting over the years has also enlightened the medical community about the effects of various hormones on every aspect of the reproduction process. Fertility patients across the world have benefited from basic reproductive endocrinological work first contributed by rodents. Once a hormone is found to be important, it is assessed in various species, with researchers ultimately working their way to humans. Of course, humans like to skip this research step on occasion. We really have no idea about the long-term effects of drugs such as birth control pills and Viagra—or even the effects of certain procedures such as C-sections and induced labor on both mother and offspring. The rodent work tells us that even minor manipulations of hormones can have dramatic effects; we should always proceed with caution.

Sexy Brains

Screenwriter Woody Allen is known for his quote about the brain: his "second favorite" organ. The truth is that the brain is just as essential to sexual behaviors as the reproductive organs. And providing the necessary fuel for the fire, the cadre of reproductive hormones—such as estrogen, testosterone, progesterone, and prolactin—is absolutely essential for pulling the appropriate neural triggers for the various reproductive responses. In rodents, if the hormones are removed, the behavior is removed; thus their sexual responses are considered hormone dependent. The critical hormones exert their effects by entering the brain through the brain's security system, the blood–brain barrier, and activating certain areas responsible for reproductive behaviors.

In the female, the focus has been on the small structure known as the hypothalamus. Generally this pinhead-size structure is involved with motivational behaviors, such as eating, drinking, copulating, escaping, and fighting. I frequently tell my students a classic biopsychology axiom that this structure is responsible for the four F's: feeding, fleeing, fighting, and . . . mating. Positioned within this small structure are clusters of similarly functioning nerve cells, known as nuclei. A specific nucleus with the technical name of the *ventromedial hypothalamic nucleus* is intimately involved with lordosis in the female rat. If this brain area is destroyed, the female will no longer display the lordosis posture required for the male to mount her. In addition, if the reproductive hormone progesterone is delivered to this area, it elicits flirting behavior in the female—you know, the hopping, darting, and ear wiggling. It is interesting that the ventromedial hypothalamus is also well known for its role in eating, especially informing us when we're full or satiated. Perhaps the female brain sees little difference between food and sex, explaining why chocolate is such a bestseller on the supposedly most romantic day of the year, Valentine's Day. Another small nucleus in the hypothalamus, known as the *medial preoptic area*, has been implicated in rodent sexual behavior. It appears that this area regulates sensitivity to circulating hormones that trigger the lordosis response. When this structure is damaged, reduced levels of hormones are necessary to exhibit the lordosis response.

The brain's reward neurochemical (dopamine) and the nucleus accumbens (a brain area previously mentioned for its role in rewarding responses) are also involved in the copulatory response. If the nucleus accumbens is damaged, female rats reject males more often than when they have an intact brain reward circuit. This complex system, made up of hormones, reproductive organs, motor systems, and key brain areas, leads to sexual motivation and copulatory responses. An interesting study in the 1970s provided strong evidence of the intensity of the female's motivation for sexual encounters. Researchers found that female rats would run across an electrified grid to gain access to a male. Thus, contrary to thoughts that females play a passive role in copulation, research continues to reveal the active role they play.

Research focusing on males reveals that there is some overlap with the female neural sex circuit. The medial preoptic area of the hypothalamus is also involved as well as an area called the amygdala, also known for its involvement in emotional processing, especially fear-related emotions. In an experiment that seemed to re-create Amsterdam's red-light district, University of Cambridge neuroscientist Barry Everitt and his colleagues trained male rats to press a bar for a sexually receptive female. After the males learned this task, the scientists damaged the medial preoptic area and placed the animals into the sex chamber. The brain-damaged rats continued to bar-press for females; however, once the female was delivered, the male rats failed to copulate with her. When the amygdala was damaged, just the opposite was observed. These males didn't bar-press for females, but if a female were presented, they would copulate with them. Thus Everitt and his team dissociated sexual desire and sexual performance, providing evidence that they are not controlled by the same brain circuits and can be differentially affected. Also similar to females, the brain's reward circuit is involved in male sexual behavior. When the male rats with the amygdala lesions, who were not interested in bar-pressing for females, had the drug amphetamine (a drug that enhances the neurochemical dopamine) infused into the brain's pleasure center, the bar-pressing for receptive females resumed. Natural increases of dopamine in the brain's pleasure center are increased when a receptive female is placed behind a screen; accordingly, dopamine in the brain's movement center, the striatum, is observed when the rat copulates.

I realize that this is a little unconventional, but I'd like to interrupt this section by suggesting that if you happen to be an adolescent male, especially an adolescent male who may date my daughters sometime in the future, you should just skip over the next three paragraphs. That's right, keep going until you see the subtitle for the next section, "Ratmatch.com." I'm sure the next few paragraphs are of little interest to you . . . don't want to waste anyone's time!

Based on the research just described, it is clear that the brain enhances sexual behavior. An interesting study conducted by Elizabeth Gould and her colleagues at Princeton University suggests that the converse is also true; that's right, sexual behavior also enhances the brain. Gould is well known for her interesting work confirming that mammalian brains produce new neurons throughout the life span. This process, known as neurogenesis, has been well documented in rats; in fact, the condition leading to the highest rates of neurogenesis is running, a phenomenon we'll discuss in Chapter 10. On the other hand, conditions associated with stress and high-stress hormones are typically associated with low levels of neurogenesis. Gould wondered if a behavior that could be described as both stressful and rewarding—namely sex—would lead to increased neurogenesis.

To test her hypothesis, the Princeton male rats were exposed to either one receptive female (acute sexual experience) or fourteen days of sexual experience (chronic sexual experience). The animals in each group were injected with bromodeoxyuridine (BrdU), a substance that is incorporated into DNA during cell division, leaving a physiological tag on recently generated brain cells; this technique is handy for discerning when new brain cells are born. The scientists also measured stress hormones and assessed anxiety behavior in the rats receiving sex therapy. The impetus for my little note to adolescent males out there was my mixed response to these data. Basically, the final report indicated that rat brains exposed to sexual behavior expressed a higher rate of new neurons, or neurogenesis in the hippocampal area of the brain, an area involved in learning, memory, and emotional processing. As I was wondering about including the study in this book, I found myself imagining adolescent males across the country downloading it to present to parents, teachers, and potential dates. Adding another interesting twist to this research sex scandal, one member of Gould's team, Erica Glasper, is a former undergraduate member of *my* team. Consequently, I am conflicted by these data. Being the parent of two adolescent daughters, I would be happy if these results never saw the light of day; however, being a former professor of one of the study's researchers, I'm extremely proud of her contributions in this fascinating study. Thus the sex therapy research at Princeton has left me with feelings of stress and reward; perhaps my brain is generating new neurons as well.

Getting back to the details of this interesting study, it is important to mention that only the acute sexual experience was determined to be stressful for the animals. The chronic sex group no longer had high-stress hormones and exhibited decreased anxiety in a behavioral anxiety test. As an added benefit, when the nerve cells were evaluated in the chronic sex group, enhanced growth of the connecting structures, known as dendrites, were observed. Thus, even though sexual behavior is stressful, at

> This study suggests that sex builds more complex brains . . . There, I said it.

least in the beginning, the rewarding aspects of the behavior appear to lead to both the production of new nerve cells and enhanced complexity of existing neurons in a brain area critical for learning and memory processes. This study suggests that sex builds more complex brains . . . There, I said it.

Ratmatch.com

I can't confidently say that rats aren't attracted to large muscles, dreamy eyes, or a big bank account. I can only confirm that the scientific literature suggests that the rats use a more sophisticated compatibility screening technique. If a female rat could write a single's ad it may read something like this:

> Female rat from City Block 8 searching for healthy young male living at least seven blocks away who isn't afraid of letting the female pace the timing of romantic encounters. Coat color doesn't matter, but a diverse immune system does—a major histocompatibility complex that is very different from my own is essential to ensure the immunological health of our offspring. I have my heart set on a family, but there's no expectation for the father to hang around. I know male rat brains aren't necessarily wired to respond to the sophisticated social interactions required for parenthood. If interested, meet me in the local alley. I only have five hours left in this estrous cycle!

I realize that most people think that very little screening occurs before rats consummate their relationships, but the truth is, at least for the females, romance is serious and shouldn't be taken lightly. For example, Greg Glass, a researcher in the Department of Molecular Microbiology and Immunology at Johns Hopkins University, has been tracking the distance traveled by the Baltimore city rats to find their mates. I was amazed to learn that his DNA testing suggests that jet-setting females are traveling a long distance—up to seven city blocks—when there are plenty of available males living in the female's local neighborhood. Why travel so far? Researchers have yet to find the answers, but one possibility may be related to a little genetic screening. Without the benefit of a genetics counselor, the female rats have to improvise—and they do this beautifully—with their impressive olfactory system.

It appears that the proteins produced by a component of the genetic immunological blueprint have a distinct odor. This part of the immune system, known as the major histocompatibility complex (MHC), is made up of a cluster of genes that encode proteins that provide information about the ongoing and potential ability of the immune system to combat various pathogens, or foreign invaders. The more diverse an animal's MHC, the better equipped it is to respond to the diverse array of immunological challenges it meets throughout its lifetime. In simpler terms, there are a lot of different germs in an animal's environment; rats with a more diverse immunological took kit have an advantage over those with more limited options. Accordingly, a female rat looks for a male whose MHC is different from her own so that her offspring will be well armed to fight disease and infection. Thus the odors of the MHC-different males are indeed sexy for the females and likely lure females from their home territories to unknown lands, or city blocks in the case of the rats. When female mice are allowed to choose their own mates in seminatural habitats, the MHCs of their offspring are more diverse than those of offspring who are the result of laboratory matings with assigned males. Thus, without the help of genetic counselors, the mice successfully screened the immune profiles of their mates, leading to increased chances of survival for their offspring.

You may be thinking that there is no way humans are using this genetic screening strategy, but it is always wise to avoid jumping to conclusions when it comes to evaluating the shared experiences between rodents and humans. A series of interesting "smelly T-shirt" studies suggests that females prefer the T-shirt odor from men less closely related to them and who have more diverse MHC profiles. Hmmm, makes you wonder what all those soaps, lotions, and perfumes are doing to our ability to choose appropriate mates. Researchers at the University of Oxford recently found that when the MHCs of couples were compared to randomly selected pairs of individuals, the couples were more MHC-dissimilar than were the random pairs. This study tells us that human couples were indeed using MHC in their mate selection, even though this certainly wasn't a conscious decision. However, with such high divorce rates these days, perhaps humans are not the model to aspire to when it comes to the romance department. In fact, I'll introduce a better model for long-term romantic relationships in the next section.

The general health of an individual may also influence his or her desirability in the animal world. One of my former students, Sabra Klein, currently a professor at Johns Hopkins University, has done extensive research on just what rodents find as sexy in potential mates. In an interesting study conducted with Ohio State University neuroscientist Randy Nelson, these researchers manipulated the health status of males from two different vole species—one monogamous (committed-relationship-type of voles) and the other polygamous (voles that liked to play the field). They designed a study, kind of a bachelorette experimental design, in which the female of each species was given a choice between two males, one that was made sick with a bacterium known as LPS and one that received a harmless saline injection. The scientists hypothesized that monogamous female voles would pay closer attention to the health status of the male since she was looking at a long future with this guy and sickness may not be a good sign about the robustness of his immune system or the length of his life. Sure enough, that is what they found. Monogamous female voles spent more time with the saline-injected male, whereas the health status didn't influence the romantic choices of the polygamous female vole.

In the 1990s, Leah Swongeur, an undergraduate student at the time, wanted to do a similar study with the monogamous California mice in our laboratory. We went to a pet store and bought cages with tunnels so that the bachelorette mouse could be placed in a middle cage with free access to tunnels leading to two separate bachelor pads with male California mice. She just had to travel through the tube to spend quality time with the guy of her choice. We put little collars on the males and tethered them to the cages so they would stay put in their own cages and allow the female to play the active role in this dating game. Our results mimicked those found in the previous study. The female California mice spent more time with the healthy mouse versus the sick one. These findings explain why we rarely see romantic epics featuring leading men with runny noses and crusty, red eyes.

Thus female rodents seem to be most interested in her mate having good genetic stock, and she uses her sensitive olfactory system to conduct her own assessment of genetic and immunological competence. When given the opportunity, her strategies often involve trying to meet a guy from out of town to avoid inbreeding and to enhance immunological diversity. Perhaps a keener sense of smell would benefit human searches for that perfect partner!

Later in the chapter, we'll consider whether rats, famous for playing the field, show any potential for extended romantic relationships.

A Prairie Vole Companion

In Garrison Keillor's radio show *A Prairie Home Companion*, he brags that the fictitious town of Lake Wobagon's inhabitants are distinctly different: "The women are strong, the men are good looking, and all the children are above average." When it comes to voles, the prairie vole also has a distinct characteristic—it forms pair bonds for life, a true committed rodent. Fidelity, however, is not high on the list for the prairie vole's close cousin, the montane vole. When given a

> Even after one dies, the partner rarely establishes a new pair bond.

choice of an existing familiar partner, the prairie voles go for the established relationship, whereas no preference is shown by the montane vole. Once prairie voles form a partner bond, the couple stays together for the rest of their lives; even after one dies, the partner rarely establishes a new pair bond. Talk about till death do us part! Could these voles be more monagomous than humans?

University of Illinois neuroscientist Sue Carter has conducted extensive research on the neurochemical basis of long-term pair bonds in prairie voles. In one study, she and her colleagues administered two neuropeptides, oxytocin and vasopressin, to male and female voles. Oxytocin is generally known for its role in lactation and childbirth but is also involved in positive social responses; vasopressin is involved in several physiological functions such as fluid retention but is also known for its role in social responses, especially with members of the same species. Carter and colleagues found that these neuro-peptides facilitated the formation of pair bonds in males and females. That is, when given a choice between a familiar partner and a strange animal, these neuropeptides were found to lead the prairie vole to spend more time with the familiar partner. When these neuropeptides were chemically blocked, no preferences for prior partners formed.

A study conducted with Karen Bales at the University of California at Davis suggests we have to be careful with dosages of oxytocin—that is, the effects don't seem to be linear. When newborn voles received a high dose of oxytocin, they were more likely to hang out with the unfamiliar male when tested later in their lives; thus high doses at certain times of development can lead to an interruption, rather than a facilitation, of pair bonds.

Hmmm, exactly how much oxytocin in the form of Pitocin did I receive when each of my daughters were induced? Maybe that explains why, as adolescents, they often choose to spend time with their less familiar friends over their familiar parents.

Tom Insel, currently the director of the National Institute of Mental Health, and Larry Young at Emory University and their many collaborators have also conducted fascinating research with these little voles. Comparing the monogamous prairie voles to the promiscuous montane voles has provided interesting clues about why these animals have such different life strategies when it comes to romance. Focusing on the vasopressin receptors that are activated when vasopressin molecules float by, Young has found distinctly different distribution maps in the two species. In the prairie vole, the clusters are centered around the lower part of the brain known for its involvement in reward. A remarkable study showed that if a vasopressin receptor gene is placed in this area of the free-loving montane vole, his commitment level raises significantly. After this research was published, the press was quick to dub the vasopressin receptor gene the "fidelity gene" and asked if women should have potential husbands tested before the wedding. Confirming the conservation of life's basic processes through evolution, recent research with human males also indicates a relationship between marital happiness and particular variations of the vasopressin receptor gene.

Young's group has also identified interesting oxytocin receptor patterns in female voles. Specifically, he has identified rich populations of oxytocin receptors around the nucleus accumbens in the brain's pleasure center as well as in the prefrontal cortex, an area responsible for more cognitive executive functions. In addition, the reward neurochemical dopamine plays a role in pair bonds in both male and female prairie voles. Thus the recipe for a romantic cocktail that has emerged from this research on these rodent Romeos and Juliets involves oxytocin and vasopressin combined with a dash of dopamine. Of course, the research suggests that this is a delicate and complex process so it's not likely the recipes will be ready for mass marketing any time soon. I am simultaneously intrigued and scared when I read about the fascinating work with oxytocin nasal spray that is supposed to make humans more trusting. Visions of spraying audiences with oxytocin at political rallies or infusing the customers with a refreshing oxytocin spray as they enter used car lots come to mind. I think I'll stick to nature's doses of these neurochemicals, being mindful that we have some control over them by monitoring our social worlds.

As I close this chapter, I confess that I feel like I've betrayed the rats by focusing all the attention in this last section on prairie voles. When it comes to romantic commitments, is there any hope for the rats? Can a promiscuous bachelor—always looking for the next best thing in female rats—ever be persuaded to develop a relationship with that special someone? I decided to dive into the scientific literature one last time to redeem these noncommittal rodents. It turns out that some Canadian rats offer a glimmer of hope that bachelors rats can be somewhat transformed. There are two necessary variables in this transformation, however—an old-fashioned game of playing hard to get and a little almond extract. When pairs are separated by a partition that allows the female to play the hard-to-get card, the male appears to experience increased arousal as he anticipates actually interacting with the female. When James

Pfaus at Concordia University and his colleagues housed rat pairs in pacing chambers, where the male had either limited or easy access to the female, the males preferred the familiar female when given a choice between her and a new date four days later; however, he preferred the familiar rat only if she wore a little almond-scented perfume. I guess this helped the commitment-impaired male rat remember their previous interlude? When the scientists wrote up their results, they didn't talk about partner preference or pair bonds as the prairie vole researchers did. Instead, this preference was harshly referred to as the formation of a conditioned ejaculatory preference (CEP). Thus it's clear that the prairie voles win the romance award for rodents; the rats are certainly reproduction and fertility experts, but their tactics are a bit more clinical and less romantic than the voles.

What the rats lack in the committed relationships area, however, the females make up for in the invested mother category. In the next chapter, we'll focus on life after sex, when the kids come along.

||||||||||||||||||||||||||||||||||||||||||||||||||||||||||||||||||||||||||||

CHAPTER 9

Family Values

In the early 1990s, Vice President Dan Quayle stirred up a national debate over family values as he attacked Murphy Brown, a character in a TV situation comedy series, for her decision to have a baby as a single mom. Quayle was clear about his position on this issue as he pointed out the dangers of portraying a successful, intelligent, professional woman "mocking the importance of fathers, by bearing a child alone, and calling it just another *life-style choice*."

As the term *family values* was being thrown around in political and media circles during this time, the many forms that families had taken on over the past half century were discussed. In a recent article about the changing landscape of families in the United States, social policy researcher Hilda Kahne at Brandeis University wrote that since the 1950s, the

percent of families consisting of married couples has decreased from approximately 90 percent with most wives classified as homemakers to about 76 percent by 2001 with half
the wives working outside of the home. The number of families characterized by individuals living with no spouse (that
is, single-parent families) has increased during the same
time, now representing about 25 percent of all families;
indeed, according to Kahne's research, single-parent families
represent the fastest-growing type of family in the United
States.

Regardless of the latest politically charged views of families
and family values, Nature has selected certain family strategies
in various species. In this chapter, we'll take a look at three different rodent families—rats, mice, and the degu, introduced
earlier. A maternal rat's brain and body experience dramatic
alterations as she prepares to care for her many helpless pups.
However, there is no evidence that comparable changes accompany fatherhood in the rats. Rat dads are not great father
figures—they prefer to leave the parenting to the rat mom.
Consequently, to see examples of true rodent fatherhood, we'll
have to focus on the California deer mouse and the degu
because both these species take their paternal responsibilities
very seriously. We will also examine rat motherhood models to
learn more about such important societal issues as raising children in poverty conditions. And, in the final section of this
chapter, we'll consider the bizarre story of bamboo fruit–
inspired reproduction adaptation exhibited by the black rats of
India. Let's start with the rats—mother rats—definitely a force
to be reckoned with.

Mother Knows Best

When I arrived at the campus of Randolph-Macon in the fall of 1989 to begin my career as a psychology professor, I was unaware that I was about to meet a longtime colleague and friend who would change the course of my research for the next several decades. I was perfectly happy with my research program that focused on stress and coping—and still am—but when I met Craig Kinsley, my counterpart on the faculty at the nearby University of Richmond, we had a lot to talk about. Kinsley had just completed a postdoctoral fellowship at Harvard University with maternal brain research pioneer Bob Bridges. Consequently, he had gained considerable experience investigating the impact of pregnancy and motherhood on the rat's brain. Having recently obtained my PhD, I was far from a sophisticated scientist at that point. I am grateful to this day that Kinsley decided to collaborate with the rookie behavioral neuroscientist just up the road at Randolph-Macon College.

Our early collaborations combined our interests in reproduction and stress. We investigated the effects of prenatal stress on subsequent stress and reproductive responses in the adult offspring. We manipulated reproductive hormones by castrating males to investigate the impact on the stress response. We found some interesting results, especially related to sex differences in stress responses. I vividly recall one long summer day of castrating Long-Evans male rats in my lab: anesthetizing the males, removing their rather large testicles, and carefully suturing their incisions. That night when I was turning off the light to go to sleep I casually mentioned to my husband that I had done so many castrations that day in the lab I could do them in my sleep; he rolled over and kindly asked me to sleep on the couch.

Although there was no shortage of research questions in our respective laboratories, our research interests shifted, rather naturally, to the rodent maternal brain as each of us began our respective *human* families. Craig was the first to have a child, and being the stereotypical sensitive guy, he often expressed his fascination with the ease with which his wife, Nancy, seemed to take on her maternal role. He started throwing around comments about how the maternal brain must go through a dramatic transformation to prepare mothers for Nature's most important role.

In the spirit of full disclosure, when I was pregnant with my first child and trying to learn everything about how to take care of my impending newborn, I worried that my own brain wouldn't transition into the maternal role. During that time I became increasingly mesmerized by the rat mothers in my laboratory. Looking at those single moms busily grooming, retrieving, and feeding up to fourteen pups, I found myself watching the rat moms every time I walked in the lab, wishing I could ask them how they knew what to do. These rats delivered their litter on their own—no doctors, no

> I found myself watching the rat moms every time I walked in the lab, wishing I could ask them how they knew what to do.

hospital, no drugs, no visiting parents and in-laws to help with meals, no husband to help with the daily tasks. Even after attending classes at the local hospital and practically memorizing *What to Expect When You're Expecting*, I still felt inadequate compared to the maternal machines I was watching day after day as they seemingly effortlessly nurtured their offspring in my lab. I was mightily impressed with these maternal superstars, and that respect increased as Kinsley and I began our research journey exploring the brains of these maternal animals.

As we embarked on our maternal brain research program, we were well aware of the many scientists before us who had described the dramatic changes occurring in the female rat brain as she prepares for the task of taking care of her young. Kinsley likes to say that mothers are not born, they are made: made by just the right alterations of several reproductive hormones, such as estrogen, progesterone, oxytocin, and dopamine; made by several structural modifications that occur in key brain areas, such as an area in the hypothalamus known as the medial preoptic area. In addition to Bridges, who is now at Tufts University, Boston College's Michael Numan has contributed a treasure trove of information about maternal brain neural circuits. Both Bridges and Numan completed their postdoctoral fellowships with the pioneer researcher of the maternal brain, Rutgers University's Jay Rosenblatt. Of course, dozens of premiere researchers have contributed to the maternal brain story, but I mention these scientists because they have been especially influential in our journey into the land of maternal research.

Although there was a lot of exciting research at that time exploring the brain areas responsible for traditional maternal behaviors, such as nursing, grooming, and nest building, we decided to take a different path. We were interested in changes that weren't necessarily thought to be specific for maternal behavior but were necessary to ensure the maternal rat could take good care of her pups. For example, the pups are helpless when they are born. About an inch long, their pink hairless bodies need the contact of their mom and litter mates to keep them warm. Their undeveloped motor system and closed eyes severely limit their ability to defend themselves. It's no surprise that the rat pups are an easy and tasty meal for other animals when the mother is off the nest. Thus it is in the rat's best interest to be very efficient when she is foraging for food. The sooner she returns to the nest, the greater the probability that her pups will avoid becoming lunch for another animal. In addition, from a metabolic perspective, lactating for fourteen is a costly endeavor that requires adequate supplies of food and a reasonably efficient metabolic rate. Energy is conserved when the stress response doesn't fly out of control unnecessarily; it should be activated only in the presence of a real threat. Although not typically thought to be characteristics of a good rat mother, we wanted to know if the brain changes leading to traditional maternal responses such as nursing also contributed to changes in other behaviors such as foraging and stress responses—ancillary maternal responses.

In our initial studies of the parental brain, we set out in each of our labs to assess spatial ability in maternal rats. Kinsley used white-furred, red-eyed, albino rats, virgins and moms, in a popular maze known as the radial arm maze. This maze requires the rat to remember the alleys she's visited as she attempts to find the one baited with a favorite food. Typically, rats are placed in the center of the apparatus and the scientist simply records which alleys each rat enters, specifying the error attempts when an unbaited alley is entered or a previously entered alley is reentered. In my lab, we used black and white Long-Evans rats in a task we adapted from another researcher, Ray Kesner at the University of Utah. We called our version of the maze the dry land maze. In this task, rats are placed in a circular arena with a diameter of about eight feet. Along the periphery, placed in familiar corncob bedding, are eight clear food wells that contain pieces of Froot Loops cereal (which, as you know by now, the rats love to eat). The rats go through training to learn which of the eight food wells is baited each day. During testing, the rats are placed at various starting points, and we measure their ability to remember how to get to the baited well. In the initial study, we ran maternal rats, virgin rats, and foster mom rats. The foster group was exposed to pups for one hour each day to sensitize them to the cues emitted from the young so they would develop maternal behaviors just as a foster parent would do.

In both labs, the rats with maternal experience performed at a superior level in the task—rat moms at the University of Richmond made fewer errors in the radial arm maze, and the biological and foster moms in my lab had faster trial times finding their Froot Loops rewards. Could it be that the pups were making the female rats smarter? These findings supported the results we were finding with our brain work: The maternal rats had increased dendritic spines, or connection points, in the brain's hippocampus, an area involved in learning and memory. The hippocampus was a relevant brain area for the investigation of foraging behavior. Excited about these results, we submitted the early brain work to present as a poster at the 1998 Society for Neuroscience meeting to be held in Los Angeles, Kinsley's hometown. We also wrote up the learning/foraging task data and submitted it as a brief report to *Nature*, a premiere science journal that we doubted would publish our rather simplistic study conducted by two professors and their gaggle of undergraduate students at small institutions.

While we were presenting the poster at the Society for Neuroscience meeting, reporter Robert Lee Hotz, currently the science writer for the *Wall Street Journal*, came strolling by just as the poster session was about to end. He stopped by our poster, asked us about our findings, and asked if he could borrow a few photomicrographs of the brains and went along his way.

I'll never forget the storm of activity I encountered when I arrived in Richmond after the conference. To my surprise, an article about our maternal rats was published on the front page of the *Los Angeles Times*: "Changes in Pregnancy May Boost Brain Power." The story continued on an inside page that included our color photomicrographs of the beefed-up neuronal processes known as dendrites as well as images of larger maternal glial cells cells known for their supportive role in the nervous system. The article even included a few comments from experts in the field who were asked to comment on our findings. I have to laugh when I think about those scientists asking the reporter to repeat the names of the scientists, especially mine, and our small institutions.

As I was driving to my office after a day of travel, I stopped at the convenience store near the college for some coffee, and staring back at me from the newspaper stand were our awkward-looking photos, obviously the only thing the media departments at our respective institutions had on file, this time on the front page of the local *Richmond-Times Dispatch*. How could all this result from a simple research poster? As I entered my office, I encountered a flood of telephone and email requests from reporters for interviews about our smart rat moms. That evening, as I sat on my bed in my pajamas with, believe it or not, one of those thick green facial masks, I received a call from the BBC and proceeded to do a professional interview as I was tucked under the covers. Later that night, my husband called from downstairs, insisting that I turn on the *Tonight Show*; sure enough Jay Leno was telling a joke about our research, claiming that researchers in Richmond had found that pregnancy makes you smarter but by the time the woman got pregnant and came to the realization that she was with a loser, it was too late—she was already pregnant. Of course, his version was funnier.

Kinsley and I were excited about the findings we presented at the conference, but we were more excited about a secret we had been keeping since that summer. *Nature* had accepted our manuscript and had agreed to publish it—but we were sworn to silence. When the *Nature* paper was published a few months later, we were hit with another wave of media attention, but we felt better about this wave since it was based on a published report of our work. We were thrilled when Pulitzer Prize–winning journalist Katherine Ellison read about our work and, after having children herself, decided to write a book, *The Mommy Brain*, about the neurobiological changes accompanying motherhood. Thinking back about the events surrounding our maternal rat research, I truly felt like the Forrest Gump of neuroscience. It is fascinating how my laboratory rats have led me to so many interesting people and opportunities, my personal version of the Lab Rat Chronicles. Most of the research that receives national attention is conducted by scientists at the most competitive universities (Harvard, Stanford, Johns Hopkins), and these researchers run their laboratories with generous budgets provided by grants from the National Institutes of Health or the National Science Foundation. I'm grateful that I currently have a generous grant to conduct research with my students, but at the time this work was presented, my lab was best described as bare bones. I had received a modest grant from the National Science Foundation to purchase a microscope and a few instruments to cut brain tissue, but the funds used to purchase supplies came from very small internal grants offered to faculty at Randolph-Macon, money that I am eternally grateful for. We certainly made the most of the resources we had and learned that the most important elements of our research program were the questions we were asking, not the cost of our laboratory equipment.

But enough about this scientific adventure; the rats are the central characters of this story. Although we appreciated the opportunity to tell the public about our initial studies, we knew these were the early stages of this program of research, and there was a lot of work ahead of us before we could fully understand the amazing transformation of the virgin rat brain into the maternal brain. We continued the brain studies, adding new groups and various brain variables in an attempt to answer our questions. My lab began a longitudinal study to explore the long-term effects of the maternal experience. To our surprise, the foraging and emotional advantages provided by motherhood continued throughout the rats' entire lives. In Kinsley's lab, older maternal rats were found to have less of a protein (amyloid plaque protein) that is characteristic of Alzheimer's disease, prompting us to think that maternal experience may offer some type of neurological protection in old age.

We also got more creative with the behavioral measures we employ and preliminary data suggest that the moms have better motor agility and strength in a series of obstacle course types of tasks. If the mom needs to return to the nest quickly during foraging excursions, she needs to be able to swiftly navigate her physical environment. The Randolph-Macon student who conducted the motor study, Torrie Higgins, also conducted an interesting study showing that the maternal rats' performance in an attention task (the attention set-shifting task was superior to the virgins' performance). This task, requiring the rats to learn salient cues in the midst of several changing cues, was used in our rat play studies and described in Chapter 6. The moms, especially second-time moms, performed brilliantly in this task, obtaining close to perfect scores. They made very few errors and, amid distractions and changing test rules, remained focused on the task at hand. We're currently excited about a series of clever studies conducted in the University of Richmond lab. Kinsley and his students exposed the rats to a cricket in a testing arena and simply recorded how long it took the rats to catch the cricket, a natural prey item for wild rats. By this time we weren't surprised to learn that the maternal rats, lactating rats in these studies, were much faster at catching the cricket, taking about 25 percent of the time it took virgins to do the job. We have replicated this cricket-hunting effect in my laboratory with the Long-Evans rats as well. These intriguing studies have sent the students back to the lab to determine the specific mechanism influencing this superstar hunting behavior—better vision, hearing, attention? You'll have to stay tuned for these results. The data are still rolling in.

Laboratory studies provide clear evidence that the female brain goes through significant transformations in preparation for her impending delivery. The media sometimes distorted our findings despite our best efforts to be conservative about our data. Being a mother does not make you an Einstein. The evidence we're collecting, along with other researchers, suggests that the brain changes in adaptive ways to enhance parental responses. There's no evidence that my ability to solve physics problems has increased since I became a mom, but there's never been any doubt in my mind about my concern for the well-being of my children and my willingness to risk everything to maintain their safety (characteristics that are indeed important for my maternal success). In contemporary terms, it's as if the brain downloads a new brain app when children come along. All of this physiological energy directed toward the maternal brain suggests that Nature has placed a premium value on mothers and their ability to raise healthy offspring. The value of fathers, beyond the insemination component of the procreative process, however, is more questionable. Nature most certainly values mothers and, at least in some cases, that appreciation has been extended to dads as well.

The *Peromyscus* Paternity Test

"But, Dr. Lambert, what about the dads?" That was a question asked by one of my undergraduate students several years ago. I was both intrigued and a little embarrassed for not asking it myself. I told the students that, first and foremost, we had to find a suitable rodent model for paternal behavior. It was clear that the rat would not be appropriate because males seldom hang around for any form of parental duties. We identified the prairie vole model discussed in the previous chapter but were discouraged as we learned how difficult it was to obtain them. Some scientists were actually trapping the voles themselves. Being a busy working mom, I knew that I had to draw the line at trapping wild prairie voles.

We soon discovered that the California deer mouse, *Peromyscus californicus*, was the ideal model for our research. It appeared that these little rodents were the poster rodents for family values. Fieldwork using DNA tests indicates that these animals are exclusively monogamous—the "until death do us part" kind of monogamy—a characteristic of human relationships that is certainly less consistent than observed in these rodents. And, more important to our research, the California mice are biparental, meaning that the males engage in all the same parental behaviors as the females except giving birth and lactating. Similar to our findings in maternal rats, initial pilot studies suggest that the males also experience foraging advantages compared to virgin animals. Related to changes in emotional responses, as discussed in Chapter 7, these mice dads were calmer in the presence of a predator odor, demonstrating fewer interrupted grooming sequences than observed in their virgin counterparts. This study suggested that the paternal mice were more on task with basic functions than were the virgin males, even when they were distracted by the odor of fox feces.

In many ways, the California mouse offers advantages over

the maternal rat model. Sure, the rats are fabulous moms, but it has been difficult to conduct research on pregnant or lactating animals. Because both of those conditions are accompanied by a plethora of hormonal and other physiological changes, it is extremely difficult to design studies with appropriate control animals. For example, when we want to assess anxiety in a lactating female, simply placing her in the testing arena is associated with many effects, such as heightened stress from being removed from pups, that are impossible to replicate in control, nonlactating females. In addition, when assessing foraging strategies, the metabolic rates in the pregnant and lactating females are altered in complex ways that are difficult to create in control animals, even if we match groups for food restriction or weight loss. For these reasons, we extended our maternal work to the postpartum phase after weaning the pups so that we could bypass these annoying, confounding variables and assess enduring effects in the moms after their physiological responses had stabilized.

I remain captivated by the maternal brain and will continue to chip away at the changes that accompany significant neurobiological transformation. The use of a paternal model, however, provides a cleaner model to assess modifications closer to the time of the birth of the offspring without the complications of the massive hormonal changes accompanying pregnancy and lactation. Hormonal alterations certainly occur in male parents but not to the extent observed in the moms.

As we contemplated the information that could be revealed by the California mice, we adopted a comparative species approach similar to the prairie and montane vole research. We compare the biparental California mice to a closely related mouse of the same genus (*Peromyscus*) that is not known for its paternal investment. These common deer mice (*Peromyscus maniculatus*) aren't likely to kill their own pups, but they are likely to exhibit indifference, a benign neglect if you will, toward their offspring. Most often we observe them to coexist in a cage with the female and pups without lifting a single paw to help the mom out. The differences in parenting strategies between these two species are dramatic. When a *P. californicus* male with previous paternal experience (that is, one that had sired a litter in the past) is presented with what we call a conspecific alien pup (an unrelated pup of the same species), he will approach the pup, pick it up, and gently groom it and then assume a crouching posture over the pup as if he were nursing the offspring. At times I get so caught up in the moment of watching these superdads that I halfway expect them to belt out a verse of "Rock-a-Bye Baby" as they gently pick the pups up with their forepaws. In fact, these dads are so similar to the moms in their nurturing responses, I have several pictures showing the pups trying to nurse from these Mr. Mom mice! In contrast, when we test the *P. maniculatus* mice with unrelated mouse pups, none of these nurturing responses are observed; they are more likely to try to attack the pup (which we monitor closely so the animals aren't injured), escape from the cage, or assume what we call a freeze response. Freeze responses are typical in high-anxiety situations such as in the presence of a predator. Freezing allows the animal to blend in with the environment or hide from the impending threat. The lack of movement decreases the probability of being spotted by another animal.

One day Catherine Franssen, a former undergraduate

student who, after receiving her doctorate in neurobiology from the University of Chicago returned to Randolph-Macon as a postdoctoral fellow, reported the results of a pilot study she had masterminded. This was a simple test: Take two *P. californicus* families and house them both in a single complex habitat, one of those with several nesting boxes and tubes running to and from all the central areas. Both families had their own housing unit in the cage and plenty of nesting material and food, kind of like living in an apartment complex with another family. This simple demonstration illustrated just how paternal the California mice are. Franssen and our research technician at the time, Zeke Hampton, were surprised to see that one *P. californicus* dad, likely the dominant dad, had pup-napped the other family's pups. The now pupless parents seemed to be a mess, pacing back and forth outside of the nesting box where their pups were being held against their will. The pup-napping male and female accommodated their little rodent hostages— grooming them and letting them nurse as if they were their own. The next day, the other parents were so stressed that they stopped eating and fell sick, seemingly weakened by the chronic stress. Accordingly, we decided to terminate the pilot study and contemplate our observations.

In many ways, this rodent family trauma was similar to the big-screen dramas depicting worried parents awaiting the conditions for the release of their precious children. Because this demonstration involves only a single case of pup-napping, it is nothing more than a case study or anecdotal evidence, but what we saw still amazes me. The *P. californicus* males are so paternal that, in this case, the supposedly more dominant male was a bully dad, stealing the other family's pups. Also interesting was the stress experienced by the pupless parents; their dedication to their pups was vividly apparent in this demonstration. Instead of retreating back to their now-quiet nesting area (hey, they had babysitters for the night), they devoted all their energy, to the point of physical exhaustion, toward their pups.

Another situation in the laboratory confirmed the strong predisposition for the California males to exhibit paternal behavior toward pups. Several years ago, when the *P. californicus* families were shipped from the University of South Carolina to my laboratory, one of the moms passed away for unknown reasons. I was interested in the effect this would have on the commitment of the dad, now that the mom was no longer around to help or, perhaps, guide the parental care. This devoted dad stayed with the pups, looking like a lactating female as he hovered over them and slept side by side with them; of course, he couldn't lactate. My youngest daughter decided to intervene and fed the pups with a dropper, returning them to the dad's nurturing attention each time. Again, this is just a single case, but it suggests that, in the absence of the mom, the dads maintain their paternal commitment.

Family Reunion

Aside from demonstrating paternal behavior, we have been interested in identifying the neurobiological events accompanying fatherhood. How do we go about investigating the activation of the paternal circuit? We created little rodent family reunions. As we thought about parental motivation, we thought that parents would be most focused on their offspring after being separated and reunited. Some of my personal parenting experiences influenced this experimental design. When I'm home with my daughters all day, I'm not very likely to run up and hug them each time I pass them in the hall, but when they have been away for a sleepover or short trip, my first response is definitely the run-and-hug response (especially when they were younger), followed by a kiss and a quick visual inspection to make sure everything is okay. Accordingly, we separated the males from their families for twenty-four hours and reunited them with their pups, recorded their behavior, and then looked at the areas of their brains that were activated during this family reunion. Both species were evaluated—the Mr. Moms and the deadbeat dads. We even threw in a foster parent group,

Both species were evaluated—the Mr. Moms and the deadbeat dads.

virgin males of each species that had been exposed to an unrelated pup for ten minutes each day to determine if they could acquire the nurturing responses. Of course, we included the standard control group—virgin males with no prior exposure to pups. The methodology is a bit more simplistic than our maternal research but challenges still existed. To control for the dads living with another animal (the mother of their pups), we housed the foster dads and the virgins in pairs with their brothers so they still benefited from living with another animal, we hoped in an affiliative, nonthreatening, positive social relationship.

Focusing on the behavior in this study, the California mice, both dads and foster dads, approached the pups, picked them up, groomed them, and crouched over them in what looked like a nursing posture. As we expected, the *P. maniculatus* males tried to attack the pups, cover them up with bedding, escape, or freeze. (Sometimes I laugh when I think about the jokes that could easily be made by extending these findings to some paternally challenged human dads.) We couldn't wait to see what was going on in the brains of these animals differentiating the Mr. Moms from the deadbeat dads. Generally, our work suggests that being a good rodent dad is the result of two neuronal events. Because the deadbeat dads' brain areas involved in the fear response are activated more than the California mice, it appears that the first step in becoming a good dad is related to the inhibition of the fear response in the presence of pups. The California mice exhibited very little activation in the brain's basic fear circuits. Once the fear response is turned off, the neuronal trigger for nurturing needs to be activated. This is where we saw some similarity with the maternal model, incorporating activation of the medial preoptic area of the hypothalamus and some restructuring in those higher learning areas not typically associated with parental responses, such as the hippocampus. Providing the neurochemical backdrop for the paternal response are the two neuropeptides described in the previous chapter: oxytocin and vasopressin. It appears that vasopressin is critical for the transformation from virgin to parent in these mouse dads, diminishing the male's fear when placed in the vicinity of a helpless pup. Oxytocin appears to be necessary for nurturing responses, such as approaching and picking up the animals to groom them.

Our explorations are still in the early phases, but it is already apparent that the paternal response is a complex neurobiological event, encompassing multiple brain areas and chemicals. I expected nothing less. Transitioning from an animal that is focused on self to an animal focused on investing resources in another animal is a huge step in the evolution of mammals, a step that probably requires a lot of brain power. Perhaps one of the most exciting findings in our paternal research is an observation in the deadbeat dads. Although in most of the brain areas the *P. californicus* dads had more vasopressin-responsive fibers than *P. maniculatus*, one component of the data suggested that there was still some room for paternal growth in the *P. maniculatus* brains. In a brain area known for its involvement in anxiety, the bed nucleus of the stria terminalis, the virgin *P. maniculatus* males had fewer vasopressin fibers, thought to be important for paternal responses, than the virgin California mice. The *P. maniculatus* with some form of paternal experience, however, either biological dads or foster dads, were not significantly different from the California paternal mice, at least in this one brain area. These results suggest that time with pups leads to brain changes that are thought to be involved with paternal experience. Considering that these animals were far from attentive fathers, perhaps the brain was beginning to lay the foundation for nurturing responses, and more time with the pups would have led to more dramatic changes. These findings make me think a little more seriously about the importance of paternity leaves that allow fathers to spend time with their young children. The findings are also in line with the research findings of James Swain, a neuroscientist at the University of Michigan, who investigates both maternal and paternal human brains. His findings suggest that, whereas the human moms' brains light up like a Christmas tree when they see pictures of their babies beginning shortly after the baby's birth, the fathers' brains take a little time to show such activation, to the tune of about sixteen weeks. Stay-at-home dads, however, show the full activation sooner than working dads. It appears that, in some cases, the paternal brain may be a bit more of a late bloomer than the maternal brain.

I'm excited to learn additional secrets about the evolution of complex social behavior from these sensitive California mice. Beyond the paternal responses, our sights are set on seemingly unique human responses, such as nurturing, sympathy, and empathy. Clues, in the form of videotapes and brain slices, are being assessed as I write these pages; there is definitely more to this fatherhood story, and I can't wait for the results!

Single Rat Moms Living Below the Poverty Line

It is important to point out that the research conducted with maternal rats and paternal mice has been conducted on animals that had all their resources provided for them in the laboratory: safe and clean housing, all the food and water the rats want, and comfortable temperatures. If we are interested in learning about human parenting from these animals, however, we need to be vigilant of the current state of the American family. As indicated earlier in this chapter, there is an increasing number of single-mother households, and the single-mom rat

THE LAB RAT CHRONICLES

model is a valuable model for the single-mom human family. Unfortunately, poverty often also goes hand in hand with such conditions. According to Kahne's review, cited earlier, more than 25 percent of female-headed families have incomes below the poverty level. Further, the number of single moms with children living in poverty was 32.5 percent in 2000, much higher than the 6 percent rate observed in households consisting of married couples with children. Low education and inadequate job skills are likely to blame for the disadvantaged circumstances of these women. Limited resources obviously lead to stress for both the mothers and children. Can rats inform us about the extent of these stressful experiences?

We'll return to some California rodents for answers—this time maternal rats at the University of California at Irvine in the laboratory of neuroscientist Tallie Baram. In one study designed to determine the extent of the effects of poverty on the moms and the quality of their maternal responses, researchers created a poverty situation by limiting the amount of bedding in the rat moms' cages. Bedding is typically made of either dried corncob pieces or cedar shavings and is placed in laboratory cages to absorb the rodents' urine. But rats often manipulate the bedding—moving it around in different patterns in the cage to serve as nesting material or a barrier from a stressful stimulus. The bedding also provides a soft barrier between their bodies and the hard Plexiglas cage bottoms and, when cold, can serve as a way to increase body temperature. In the poverty experimental group, instead of having a standard layer of bedding, the cages were fitted with a soft plastic bottom that would catch the droppings. In addition, each mom was given a single paper towel for nesting material. These conditions were stressful, but perhaps not to the degree of some single human moms face with both limited resources and increased violence threatening the well-being of her children. The control group in this study was given the standard amount of animal bedding. The moms were exposed to these conditions from postnatal days two through nine, a mere week of limited resources. On the stress scale, this is most likely in the low/moderate category. Would one week of limited resources have a significant effect on these rodent families?

Baram and her colleagues found that maternal rats living in poverty were profoundly affected, all negatively, and the stressed moms likely compromised the development of their offspring. Observations of the quality of maternal care indicated that the poverty moms exhibited less licking and grooming; further, they spent more time off the nest than the control rat moms. The moms exhibited increased anxiety in the form of less exploratory behavior in the open field test. The stress hormone corticosterone was increased in the poverty moms, and the adrenal glands (the glands atop the kidneys that secrete stress hormones) were heavier, indicating overactivity, in the moms living below the laboratory poverty line.

Remember the research conducted by Michael Meaney and his colleagues described in Chapter 7, suggesting that normal maternal rats with naturally low-licking rates compromised the neural and emotional development of their offspring? In light of those results, the findings in the poverty study, with its more dramatic maternal interruptions, are disconcerting. What impact would the fragmented maternal care in the poverty moms have on the development of the pups? Baram's team addressed that question by following the development of the pups into adulthood. It is interesting that, whereas no learning or cognitive differences were observed in early adulthood (four to five months of age), cognitive decline was noted in middle-aged rats (twelve months of age). In one task known as the Morris water maze, rats learn that they can escape from forced swimming in a pool by finding an immersed platform that remains in the same location in the swim arena. Learning is determined by recording how fast the rat can swim to the platform, the location of which had been learned in previous trials. Focusing on the hippocampus, which is involved in learning and memory, researchers observed both structural and physiological compromises in the rats raised in poverty for a single week. Specifically, dendrites, processes on hippocampal neurons that serve as the basis for neural connections, were less extensive and complex in the rats raised in poverty conditions. These results confirmed that one week of disrupted parental care led to an accelerated decline of key areas of the brain involved in learning and memory, functions critical for adaptive responses and survival.

Similar negative effects have also been observed with disrupted paternal care in degus, the South American rodents briefly introduced in Chapter 7. German neuroscientist Katharina Braun and her colleagues conducted a microscope analysis of the orbitofrontal cortex, a chunk of the highly evolved cerebral cortex known to be involved in social interactions. They were interested in the cellular complexity in degus either raised by both parents or raised by a single mom. Removing the father from the picture in this species that has evolved to depend on the dad's help resulted in delayed maturation of the orbitofrontal cortex—in this case, it appeared that the moms were not able to compensate for the absence of the father. Childcare in this species is definitely a two-rodent job.

Thinking about the long durations of extreme stress experienced by both children and their parents in our society, rats are revealing that the search for ways to diminish these early-life stress conditions should be a high priority. Compromised brain development leads to compromised functioning as well as an enhanced vulnerability to mental illnesses such as depression and neurological conditions such as Alzhemier's disease. The delayed maturation in the orbitofrontal cortex, if generalized to humans, could result in altered social interactions, perhaps predisposing children toward symptoms related to autism or social anxiety. The provision of safe homes for our children appears to be a wise investment in both short-term and long-term health. Accordingly, policies designed to enhance the quality of childcare appear to be an excellent investment for healthy brains—for rodent pups and human children alike.

The Mother Load

In northeast India, wildlife biologist Ken Aplin recently explored a rat family mystery of sorts. It has been reported that every forty-eight years a plague of rats invade rice fields, wiping out crops that farmers and their families rely on for food and income. Another environmental event is documented to occur on the same schedule: the flowering of the bamboo plants that are so prevalent in this area. This flower leads to the massive production of a fruit, and the vast quantities of this fruit provide nutrients for black rats, previously observed to be struggling to secure resources to maintain their pregnancies and offspring. Aplin, accompanied by a *National Geographic* film crew, traveled to India to witness the next round of flowering bamboo and rat plagues, known as *mautam* in India. He hypothesized that the opportunistic black roof rats were taking advantage of this new and plentiful food resource and, accordingly, were able to ramp up their reproductive rates to phenomenal proportions, yielding the true mother load of rat pups. His research in the village of Mizoram, surrounded by twenty-four hundred square miles of bamboo plants, confirmed his hypothesis. Based on his calculations, the hundred rats existing before the plagues became twelve thousand rats after four waves of reproduction (occurring in less than six months) for the original females. Aplin based his predictions on the number of implantation scars in the uterus of captured females. Of course, these bamboo fruit riches don't last forever—forcing the new army of rats to search for other food options; the rice fields are often next on their list. Within a single night, an entire rice paddy can be demolished, completely stripped of the precious rice needed to sustain the human families over the next year. Once the rice fields are cleared, the bulk of the rat surplus dies, returning the population to the meager levels existing before the bamboo fruit harvest.

The ability to take advantage of changing conditions in a dynamic environment is a priceless neural commodity in the evolution game. The maternal rats in India confirm these rodents' status among the most impressive mammalian opportunists. Regardless of numbers of offspring, however, the parental response itself is rather revolutionary in the grand scheme of evolution. According to the great neuroanatomist and theorist Paul MacLean, who spent most of his career conducting research at the National Institute of Mental Health, the onset of parenting was a critical time for the evolution of the mammalian brain, the mother load of evolutionary opportunities for these animals. In his landmark book *The Triune Brain*, MacLean wrote that the maternal brain guided the evolution of complex mammalian brains. Compared to reptiles, MacLean pointed to three critical behaviors leading to the complex mammalian brain, all related to early family life: nursing, audiovocal calls, and play. With increased opportunities for social interactions, communication, and the development of a brain with a neuron-dense cortex, the emergence of the family on the evolutionary scene was indeed the mother load, ultimately leading to a species that can build cities, generate art in multiple forms, and strategize about the well-being of future generations. Family life, and all the complexities associated with it, allowed us to leave our egg-dropping reptile cousins behind and move on to more advanced neural networks and all the advantages and disadvantages that come with them.

In this chapter, I have described research questions that keep me up at night, that fuel passionate conversations in my laboratory, and that help me understand that the social interactions in my own family with my husband and two daughters have left the most significant footprints on my neural structures. As the form of families take on new characteristics, the rats remind us of a few tips to keep in mind. First, having a child is a big deal for the brains of mothers. Although human moms often report being scatterbrained, Nature has provided neural insurance to guarantee that the appropriate attention is directed to the offspring. The maternal brain isn't fail-proof, however; stressful conditions and limited resources can affect maternal care, resulting in long-term effects on the development of the offspring. Second, at least in some species, fathers matter. They contribute significantly to the care of the offspring, and their removal results in compromised brain development. Finally, and perhaps most important, the fundamental building blocks for all social relationships are likely influenced by the neural circuits that enhance familial responding. You may have noticed overlap between the neural circuits and chemicals associated with the sexual behavior discussed in the last chapter and the parenting responses. And these circuits likely extend to friendships and other caring relationships, perhaps serving as the basis for sympathy and empathy, traits that are seldom extended to nonhuman mammals. Indeed, our family ties may run deeper than we thought. Tracing the ancestral roots in rodent models may teach us a thing or two about contemporary human families as well as provide insights into the development of higher-order cognitive endeavors such as the emergence of our human social conscience.

||||||||||||||||||||||||||||||||||||||||||||||||||||||||||||||||||||||||||

CHAPTER 10

Winning the Rat Race

Rat races conjure up images of being trapped on never-ending treadmills racing to some unknown destination, exhausted from sleep deprivation and chronic stress. In my experience with both humans and rats, humans are more likely to find themselves in these conditions than are rats. At least that's the case for the wild rats, for whom energy is a valuable commodity. Field trips from the home burrow are reserved for meaningful purposes, such as locating mates, food, or shelter. Running around aimlessly is likely to put the rat on the fast track to being lunch for predators.

Running in an unnatural environment is a different story. Context is incredibly important when it comes to lab rat races. Laboratory studies have investigated several forms of running conditions. Rats and mice have been housed in cages with adjoining wheels and allowed to run whenever they want. And the variations of these running rodents continue: They have been placed on a diet while they were housed in cages with activity wheels, housed in cages with a running wheel that was locked so that the rats couldn't actually run, and housed in cages with no running wheel but forced to run on a treadmill for about an hour a day. The results in these various conditions are dramatically different, resulting in several potential outcomes such as (1) a burst of new brain cells and new blood vessels pumping resource-rich blood to the brain areas, (2) a burst of new stem cells that won't live long enough to develop into mature brain cells, and if there is no intervention, or (3) no evidence of brain benefits and death within about five days. Can you guess which groups produce the specific outcomes? If not, read on . . . the answers are no doubt much more important than winning a race; they are a matter of life or death. Before jumping into the brain benefits of running, we'll consider the drastic changes in neuroscientists' views about changing neural landscapes over the past century. New laboratory technologies have illuminated surprising truths about the rats' neural cell populations, and these observations are revolutionizing views about maintaining human brains.

Brain Stains

Less than half a century ago, students were taught that mammals are born with all the brain cells (known as neurons) they would ever have throughout their lifetimes. The take-home message was simple, to hold on to what you were given—wear the bike helmets, avoid toxins, protect your neurons any way you could. Over the past several decades, this neuronal dogma started to change as researchers confirmed that new neurons were indeed produced throughout a mammal's lifetime. The production of new neurons, or neurogenesis, is easily confirmed if you inject bromodeoxyuridine (BrdU)—the synthetic DNA analog we talked about in Chapter 8—into a rat before a particular experimental manipulation (such as beginning a running training program). This drug is unique because it is incorporated into new cells formed by cell division. After the rat brains are collected, the tissue is exposed to a tedious protocol of antibodies and various washes, then sliced and mounted on microscope slides. If all goes according to plan, the tissue mounted under the coverslips on the slides provide clear evidence of new neurons in the form of darkly stained cells that were generated on the day of the BrdU injection. At least this is the case if the scientist waits the appropriate amount of time between the BrdU injection and the harvesting of the brain tissue. How much time does it take to grow a new neuron? About four weeks is adequate for a mature neuron; however, immature neurons can be detected after about a week.

In 1998, Rusty Gage, a noted neurogenesis researcher at the Salk Institute, and his colleagues got the world's attention when they confirmed that humans also exhibited evidence of neurogenesis throughout their lives. Gage took advantage of the fact that BrdU was used as a form of cancer therapy and wondered if the patients' brains would show neurogenesis like the laboratory rats had shown after receiving injections. In this bittersweet study, the brains of the cancer patients were autopsied, and the results confirmed the staining of the new cells. Neuroscience owes so much to these brave people who donated their brains to this investigation. Solid evidence now suggested that the rodent models were indeed a good model for human cellular proliferation and neurogenesis across the life span.

Merely establishing that a newly divided cell is present in the brain is interesting, but most researchers want to know the type of cell formed. The cells can be either a neuron or a glial cell (a nonneural cell that can serve many functions such as the delivery of nutrients to neurons and the removal of toxic wastes). To determine the cellular profile, a process known as *immunocytochemistry*, a technique that uses antibodies to target specific proteins of interest, is used. In many cases, the brain tissue is double-stained for both the BrdU antibody mentioned earlier and another antibody to convey that the cell is either a neuron (neuronal nuclei [NeuN] stain) or a glial cell (glial fibrillary acidic protein [GFAP]). Another antibody that marks immature neurons, known as doublecortin (DCX), can be used as well. Once the brain tissue is sufficiently prepared for analysis, scientists scan the brain for cells that are both black (indicating the cell was born the day of the BrdU injection) and brown (indicating the neuronal or glial profile) so they can confidently confirm neurogenesis (the creation of new neurons) or gliogenesis (the creation of new glial cells). Timing is an important part of neuronal profiling; we often use Ki-67, an antibody that marks newly formed undifferentiated cells, or neural stem cells. It is interesting that there are limited brain areas where neurogenesis takes place. In laboratory rats, these neurogenic zones are localized in the hippocampus (involved in spatial learning) and the olfactory bulb. Other areas, such as the neocortex, have been implicated as well; it is likely that the list of neurogenic zones will continue to grow as more and more research is conducted in this area.

As you might guess, all of these antibodies and histological (tissue examination) protocols require a lot of expertise and time. Because I'm not a chemist, I rely on ready-made protocols and, each time we run brains through these two- to three-day procedures, we hold our breath as we wait for the results. I never take this process for granted or feel confident that it will work 100 percent of the time. If just one step out of the forty or so is missed or altered, the whole process can be a flop. The stakes are high. If the lab procedure is not perfect, the information is lost: months and years of work, thousands of dollars, gone in an instant. As I often declare to my students, these tedious staining procedures could be a recipe for obsessive-compulsive disorder. We become obsessed with checking and rechecking the steps that must be followed precisely, and we are simultaneously plagued with fear and doubt until the thin little brain slices are safely mounted on microscope slides so they can be painstakingly quantified and analyzed.

As a student of behavioral neuroscience, I have found it exhilarating to have the opportunity to observe neurogenesis research gaining increasing popularity in the scientific literature. With so many double- and triple-labeled cells, scientists are now using what is known as fluorescent stains, which means the stained tissue will jump out in the form of fluorescent images against a dark background. These fluorescent cells are stunning: bright red neurons with a yellow center organized in a pattern that holds clues about the basic building blocks of our memories, hopes, desires, fears. One day I want to host an art showing featuring these beautiful double- and triple-labeled cells. I can see the titles of the displays as I write these words: Maternal bonding in green BrdU and yellow NeuN. Social isolation in blue BrdU and red GFAP. Love in red Ki-67 and blue DCX.

To keep from getting too overwhelmed with the histological procedures, I sometimes picture Martha Stewart hosting a "do it yourself immunocytochemistry" segment on her television show. I envision her telling the viewers that they are going to make gorgeous blue and red double-fluorescent hippocampal sections. Then she reads off the ingredients that are organized on her kitchen countertop. "For today's project we'll need forty trained perfused rat brains, a cryostat to cut the brains at just the *perfect* thickness (forty microns), and the various primary and secondary antibodies and other washes and chemicals for staining the brains at that *perfect* intensity." Then she would pull out a tray of previously stained microscope slides from below the counter and pop one under the microscope lens, prompting the audience to gasp at the wonder of the beauty of the cells. "Immunocytochemical detection of neural plasticity . . . it's a good thing."

Running for Neurons

What has research revealed about the impact of running on neuron production? Several studies have been conducted on rodents, especially mice in this case, showing strong evidence that having access to a running wheel results in increased rates of neurogenesis in a specific area of the hippocampus known as the dentate gyrus. In fact, these studies suggest that voluntary running can produce up to a fourfold increase in neurogenesis. Let me say that again: Running leads to a fourfold increase in the production of new neural cells. If we can establish that the new cells are actually functional in adult mammals, we're looking at a potent neural stimulator that requires no drugs,

Running leads to a fourfold increase in the production of new neural cells.

no surgery, no high-tech equipment, no expensive treatments, no insurance forms. Wow.

Peter Clark and his colleagues at the University of Illinois at Urbana-Champaign showed that the existing cells in the hippocampus were more intensely activated during times of peak running than at other times. The running logs for this study indicated that the running levels in mice reached a peak after about twenty days and then maintained a plateau. For some reason, the hippocampus really pays attention to the running rates of these rodents. Compared to sedentary animals, twenty-five days of running nearly doubled the chances that new cells would eventually mature into neurons.

The running-induced neurogenesis effect is so robust that hippocampal neurons impaired by irradiation can be rescued in running animals. Knowing that cranial radiation, used for nervous system tumors, has a side effect of learning deficits later in a patient's life, researchers have delivered this treatment to mice so they could determine the impact of both running and irradiation on neurogenesis. After irradiating the mice at nine days of age, animals were assigned to either a running or sedentary group; three months later, their brains were examined for evidence of neurogenesis. Irradiation decreased neurogenesis rates, but running restored the proliferation rates in the irradiated animals to baseline levels. In addition, researchers observed that the extensions of the new neurons were distorted as they left the cell body, an effect also reversed by running. Further, behavioral modifications observed in the irradiated animals—namely anxiety behavior in an open arena—showed improvement in the running animals.

In a similar study, some Canadian rats revealed further interesting data providing additional clues about the role of neurogenesis in learning. In this study, irradiation was delivered to one group of runners and one inactive group, called the sitters; in addition, comparable nonirradiated groups were included. Once again, irradiation decreased the number of new neurons, whereas running increased them. In this learning task, rats were trained to associate an environmental cue, in this case a tone, with the subsequent presentation of a brief shock. After a while, the animals exhibited a freeze response when the tone was presented. The freeze response represented learning because the animals learned that the tone predicted the presentation of the shock. In this artificial situation, one in which escape is not a possibility, the freeze response is typical, indicating the animal's fear of the impending shock. Past research indicates that the longer the duration of the freezing response, the stronger the memory. Accordingly, reduced neurogenesis was associated with shorter freeze durations in this particular memory task. However, a lack of differences among groups in another study that used a spatial memory task reminds us how complex these questions are; at this time we must conclude that many questions remain about the function of adult neurogenesis.

A similar rescue effect was observed in a special breed of

mice that are neurogenesis-challenged in the sense that they have low levels of baseline, day-to-day neurogenesis rates. These genetically engineered, or transgenic, mice are known as syn-Ras mice because they have altered patterns of a protein (Ras) known for its involvement in neural plasticity. Having free access to a running wheel reversed the genetic-based neurogenesis deficits in these mice and, as an added bonus, even enhanced the complexity of the processes, known as dendrites, extending from the cell bodies. When tested in a learning paradigm, running enhanced the memory of both the transgenic and the standard mice. As an explanation for potential mechanisms, researchers at Germany's Ruhr University suggested that the neural growth factor known as brain-derived neurotrophic factor (BDNF) was positively correlated with running levels. This is a wonderful example of behavior redefining genetic destiny, an excellent case of epigenetics (as discussed in Chapter 7).

The positive neural effects of running are not specific to voluntary exercise. Yong-Seok Jee and his colleagues in Korea designed a study in which mice were forced to run on a motorized treadmill for thirty minutes a day for six weeks. Increased rates of neurogenesis in the running animals were associated with improved short-term and spatial memory in both young and old rats. The effects observed in the older rats were especially interesting, considering that they were showing evidence of cognitive decline before their running therapy. Running also suppressed the death rate of existing neurons. These results suggest that you can teach an old brain a new trick or two . . . and it still responds.

Thus running, even when it's not a product of our own

volition, leads to positive effects on the brain. As mentioned, in addition to neurogenesis, other positive factors such as increased brain growth factors, reduced rates of programmed cellular death (apoptosis), straight and orderly processes extending from the neurons, decreased anxiety responses, and enhanced learning in several tests are among the many perks to running rat races.

But wait—there's more! Thirty days of wheel running has been found to increase angiogenesis, the creation of blood vessels, increasing blood flow to the motor cortex in the rat brain; a mere ten days of wheel exposure enhanced blood flow in the hippocampus in mice. Blood carries valued nutrients, such as oxygen and glucose, to energy-hungry neurons. Thus increased blood flow provides more resources for the neurons to conduct their business.

When researchers throw one more type of movement into the mix, some acrobatic training involving a bit more thought or motor learning, there is yet another brain benefit. With this condition, rats exhibit enhanced connection points, known as synapses, among existing neurons—more integrated neural circuits, if you will. University of Illinois neuroscientist William Greenough, well known for his long career of exploring the effects of environmental enrichment on the brain, has created a rat version of Cirque du Soleil in his laboratory, complete with a rope ladder, a high stepping apparatus, and little rat seesaws. The culmination of all these brain perks seems almost too good to be true.

Typical of all such miraculous results, there are some caveats to this brain phenom. Thomas Hauser's team in Switzerland trapped wild long-tailed wood mice and evaluated the impact of both running and environmental enrichment on neurogenesis levels in these animals. Whereas both variables are associated with neurogenesis in their laboratory-bred mouse colleagues, the wild animals weren't affected by these so-called enriched conditions. Uh-oh. This brings up some critical questions about the validity of the laboratory rodent models. The authors of the study suggest that domestication, known to shrink the hippocampus, the hotbed for all this activity, could result in a neurogenesis compensatory effect, increased neurogenesis rates in this case, in the lab animals. Further, the wild rats run a lot in their natural environments, an effect that may have set their neurogenesis rates rather high compared to the laboratory rodents; if so, the exposure to running in the lab after their capture may not have altered their already revved-up neurogenesis rates. Although we need to remain cautious about some of the lessons learned from laboratory rodents, I provide some evidence in Chapter 12 confirming that lab rats haven't lost all of their wild ancestry.

A second caveat is related to the specific housing conditions of the rats—whether they have a support group to return to each day. Liz Gould's team at Princeton University, the same group that investigated the effects of copulatory responses on neurogenesis discussed in Chapter 8, reported that transferring rats from standard group housing in their lab to individual housing suppressed the positive effects of running on neurogenesis. They related the lack of neurogenesis in the isolated rats to an increase in stress and stress hormones. Just as the intake of food can influence the metabolism and effectiveness of various types of medicine, an animal's stress response influences the positive effects of running on neural plasticity.

What about the locked wheel condition described earlier in this chapter? Surely just looking at a wheel can't promote all these positive neural effects. Well, the answer is a bit complicated. Karl Fernandes, a neuroscientist at the University of Montreal, and his colleagues conducted a very clever study to answer this question. They exposed rats to a cage with a running wheel, but they added an interesting control group—animals that were housed in a cage with a *locked* running wheel, so the rats couldn't run. I chuckled when I read the author's description of the study: "We investigated the possibility that running wheels might promote neurogenesis, in part, through mechanisms that are independent of running." When they looked at basic cell proliferation, long before those cells differentiated into neurons, the locked and running wheel groups had comparable numbers of new cells, and both groups were higher than the control group, a group that had no wheel at all in the cage. I was intrigued by these results; did this mean I could just *look* at my treadmill and generate new brain cells? This is a great finding.

Hold on, though. Before you throw out your sneakers you need to know the rest of this research story. When the scientists looked later in the neural trajectory time line, they found that the locked wheel group's neurons were less likely to actually develop into mature neurons. They also noted that the laboratory mice did far more than just look at their locked running wheel. They climbed on the screen, hid underneath, and moved bedding around it; the mere presence of the wheel promoted more physical activity. So, to be comparable, you would need to hang from the handles of your treadmill, climb on top of it, and try to scoot beneath it each day to enjoy that early cell proliferation effect. When Fernandes emailed me about this, he explained that looking at your treadmill turns on and revs up the brain's neurogenesis machinery, but the new neurons can't mature and get integrated unless you actually hop on and start running. On second thought, we need to just suck it up and run on our treadmills, preferably throwing in a few acrobatic tricks here and there just to maximize the neural effects.

The Dark Side of Running: A Cold Case from the Rat Files

When I was a second-year graduate student I had an opportunity to take a seminar course from my research mentor, Lee Peacock, along with five or six other biopsychology graduate students. The topic was biological aspects of motivation. I was simultaneously excited and scared. I was fascinated with the topic, which addressed the mysteries of why we do the things we do. On the other hand, this was my first seminar course with my research mentor, there was a lot of pressure to perform well, and, adding to the pressure, there was an expectation that I would locate interesting research articles that would lead to my doctoral research. I had conducted the research for my master's thesis, a study exploring the effects of an exercise training program on stress responses, using humans as subjects. With all due respect to the human readers of this book, I quickly knew that I never wanted to do that again. My entire project was at the mercy of whether I could persuade undergraduate college students to come back to the lab for their posttest measure. Obviously not something that was high on a coed's list of things to do while attending a university. I called, left messages, called, left messages . . . and barely got enough students to complete the study. Unlike laboratory rats, people all had their reasons—those dreaded frontal cortex–inspired cover stories—for why they had missed scheduled appointment after scheduled appointment. My professional interests turned to the rats, and I have never looked back!

As I combed the journal stacks at the University of Georgia's science library, I felt overwhelmed by all the brain science around me. How did you know when a study was the right one (that is, one that would keep me interested for years to come, lead to a fulfilling research career, and translate into something meaningful for humans)? Perhaps I should have consulted a research program online match service (eResearchHarmony.com?). I was serious about finding that perfect line of research, and I looked up keywords in huge periodical indexes and randomly pulled out bound collections of behavioral neuroscience journals. While sitting on the nearest stepping stool, I flipped through the pages searching for appealing titles and abstracts. Thinking back, this all seems so archaic. Today I sit at my desk and access a lot more literature in an hour using electronic search engines than I could in an entire day of actually visiting the library. Although I wasn't carrying around a club or using a bone to pull my hair back in my early graduate school days, I was certainly operating in a pretechnology age.

As fate would have it, I found *the one*—the study that got my attention and has kept it to this day. It presented a neurobiological mystery, originally written by the grand author Nature herself. Let me present the facts.

The study that caught my attention was published in the journal *Physiology and Behavior* in 1976. The investigator in the case was William Pare, a scientist working at the VA Hospital in Perrypoint, Maryland. The subjects were laboratory rats, and the experimental conditions were quite simple: They involved housing rats in a cage with free access to a running wheel. An additional twist was that they had access to food for one hour per day. The rats could consume all the food they wanted in that hour, so it wasn't terribly restrictive. I remember thinking as I read through the protocol that it all sounded pretty innocuous; how bad could it be? The answer was revealed later in the article.

Within about five days of these conditions, most of the rats will literally run themselves to death. When the rats were sacrificed and dissected, they were peppered with severe gastric lesions, the stress ulcers Hans Selye had described, suggesting that this was a stressful situation. Their behavior was puzzling because instead of decreasing their activity in the running wheel due to the limited food resources, most increased their running to excessively high levels (up to five miles a day in some cases). And it was surprising that when food was presented, they didn't eat voraciously. At times, they would jump in their wheels and continue to run instead of hopping on their food dish and eating. This activity–stress paradigm was used as a stress model to produce stress ulcers, no real mystery there: The stress was causing, or at least contributing to, the gastric ulcers. I was most intrigued, however, by the question of why these rats, typically so logical and adaptive, were screwing this one up so royally. When food is depleted, it seemed that the adaptive response would be to slow down and conserve energy and take advantage of food resources when they were made available. What were these rats thinking?

After presenting my article on this activity–stress paradigm to my mentor and fellow students in the seminar class, I was officially hooked on this research question. I immediately started planning my dissertation, exploring the effects of the neurotransmitter serotonin on the responses and ulcers. This was a time before the gastric ulcer world had been revolutionized by the discovery of the bacterium *Helicobacter pylori* by Nobel laureate and Australian physician Barry Marshall; consequently, my mentors were excited about trying to identify variables that would minimize ulcers. My dissertation research was mildly interesting, enough to earn my doctorate, but the data didn't satisfy my curiosity about what was driving this bizarre behavior. If I could solve this mystery, perhaps the results would apply to a condition in which humans exhibited almost exactly the same behavior. The behaviors observed in the rats included increased activity, decreased food consumption, and extreme weight loss. Sound familiar? Patients diagnosed with the eating disorder anorexia nervosa exhibit the same behaviors: excessive activity, decreased food consumption, and extreme weight loss. I realize that the nature of the original triggers for the eating disorder are probably very different from the activity–stress paradigm—I don't think the rats ever looked in a mirror and declared that they were too fat—but symptom for symptom, this is a valuable animal model for anorexia.

After beginning my career at Randolph-Macon, I collaborated with a biopsychologist at Virginia Commonwealth University, Joe Porter. Porter was interested in the role of dopamine in motivation, and I convinced him it would be interesting to consider the role of the brain's reward neurochemical dopamine in the activity–stress paradigm. We decided to deliver a drug that would block the dopamine receptors in the animals' brains without affecting the rats' motor abilities. If running was leading to a dopamine hit, then blocking the dopamine, and perceived rewards, should result in less running in these rats. We found that the drug pimozide did not completely wipe out the excessive running, but it did drastically reduce the levels. Thus it appeared that, once the animals started running, the brain dopamine levels played a role in sustaining the behavior. But why did the rats start running in the first place?

A clue to this question came to me before the dopamine study; it just started making more sense the more research I conducted on this topic. During the summer I began my doctoral research, there was an air-conditioning problem on the sixth floor of the psychology building, the very floor where the animals were housed. Athens, Georgia, in July was a hot place . . . this was not good. I moved all the fans I could into the area and monitored the temperature closely to make sure it didn't get dangerously high. And I monitored the activity of the rats in the running wheel. Or at least I tried to, but I started to panic as I realized that the animals weren't running. Apparently room temperature had something to do with this particular form of a rat race. Oh no! No running, no ulcers, no data, no dissertation.

After I had my data pity party, I finally realized the value of what I was observing. I had to modify my research plans to accommodate this temperature snafu, but now I knew that temperature was an important variable. Could it be that in normal lab temperature climates restricted food led to reduced body fat and a diminished capacity to maintain body heat? Increasing physical activity in the form of running increased body temperature. Once the animals started to run to maintain comfortable body temperatures, the movement likely increased dopamine levels in the brain, a rat version of runner's high. These brain rewards were more powerful than the rewards associated with consuming food, leading to a maladaptive response of increasing activity in the face of restricted food. In a subsequent study, I manipulated ambient temperature and confirmed that the rats ran more in the cooler temperature condition.

When I put myself in the place of the hungry rats, I began thinking about yet another piece to this behavior puzzle. What if the running wheel activated a feral type of response, and the animals were trying to run to a different area to find food, just as they would in the wild? Of course, they didn't get far with a running wheel, but it could be that the conditions of the activity–stress paradigm flipped a feral trigger in these laboratory rats, calling back to some predisposed responses that likely led to the survival of their wild rat ancestors. If a patch of ground, previously occupied with food, became barren, it was smart to move on in search of new resources.

Thinking back, this theory explains an odd occurrence I observed when my new laboratory was under construction. I had activity wheels and a few cages of rats in another laboratory room that we were using in the transition. Because we didn't have any rats housed in activity wheels, I was surprised when I walked in the lab one morning and heard the faint squeaks of a running wheel turning. What was going on? Had a rat escaped from its cage on the shelves across the room and jumped in a wheel? I located the turning wheel and was shocked to see an animal running at top speed. But this animal was no rat—it was a squirrel. I looked up at the ceiling tiles and quickly located the point of entry. This squirrel had chewed through the ceiling tile, probably feasted on the rodent chow easily accessible on top of the rat cage lids, then decided to get the heck out of there—by jumping into the running wheel!

After these investigations, I can't claim that I have solved this fascinating rat behavior cold case, but I have at least uncovered several possible suspects, and it is likely they all have a hand in this complicated neural heist. We're probably talking about a convergence of several variables acting on the animals' brain to generate a response that, while it may be adaptive in the wild, is clearly maladaptive in the laboratory.

Before moving on to the final section of this chapter, it is important to consider another intriguing aspect of the activity–stress paradigm. Why exactly is it so stressful? The rats get a little cold and hungry and try to travel to a new place to locate food; why the crazed, stressed, life-threatening persistence? Why not adapt as they are so famous for doing and jump out of the wheel and start eating when food is presented to them? What caused the rats to lose contact with reality? I think the answers to these questions involve a psychological aspect of the paradigm, the perception of the loss of control. The lack of a connection between effort (running) and rewards (food) creates a very stressful environment for these animals.

Special Forces Rodents

Several years ago, I had an opportunity to visit the Walter Reed Army Institute of Research. I still remember walking down those old, cavernous, dark halls, thinking back to the days when several of the pioneers of behavioral and clinical neuro-science conducted groundbreaking research on the premises. Research was still being conducted, however, and one line of studies definitely got my attention. Following that visit, I main-tained contact with two neuroscientists, Jean Kant and Sally Anderson, about a chronic stress paradigm they had been investigating. In an attempt to learn more about the nature of chronic stress experienced by soldiers, they developed a model that, when described to me, seemed like it would definitely throw the rats off their game, likely leading to a rodent version of a mental breakdown. Rats were initially trained in a classical conditioning regime in which they learned that if they pulled a little ring dangling from the ceiling of their cage when a light was presented, they could avoid the administration of a mild shock that would follow the light stimulus. This seemed reason-able, but then the chronic stress began. The rats were presented with the light every five minutes, twenty-four hours a day, seven days a week for up to two weeks. This seemed unimaginably stressful to me—not being able to sleep longer than four min-utes, or rest or focus on anything without the fear of being shocked if vigilance wasn't sustained. These conditions simu-lated conditions soldiers found themselves in at various times, especially while occupying hostile territories. But here's the surprise: When the researchers measured the stress hormone corticosterone, the levels were no higher in the chronic stress group than in the control group.

In my conversations with Kant and Anderson, I expressed my confusion with this rat research. I explained that in the activity–stress paradigm—an experimental condition in which the animals could do whatever they wanted, rest or run; they just needed to eat all they could in one to two hours each day—the animals were incredibly stressed. As mentioned earlier, these animals were plagued by high stress hormone levels, gastric ulcers, swollen adrenal glands, and shriveled-up immune glands. Yet the severity of stress in the activity–stress paradigm seemed like nothing compared to the conditioned stress paradigm developed by the Walter Reed researchers. Even more mind boggling, the Walter Reed rats learned to maintain the same duration of sleep as the control rats (even though they had to wake up every five minutes). These rats seemed to be rodent versions of special operations soldiers.

We identified one critical difference between the two stress protocols. There was one thing that the Walter Reed rats had that my activity–stress rats didn't—the psychological edge provided by a sense of control. No matter how annoyed the Walter Reed rats were with their chronic stress paradigm, they always had control over the situation because, if they followed through with the correct response, they could avoid being shocked. Every time they pulled the ring, they succeeded in blocking the shock: mission accomplished, time after time after time. In contrast, regardless of the distance traveled in the activity wheels, no food was presented in the activity–stress rats. The rats had no sense of control, leading to a life-threatening level of stress. This realization motivated me to search for ways to increase that important perception of control (such as the effort-driven rewards) in our laboratory rats.

Thus far, I have provided evidence that running can be a powerful therapeutic, brain-building behavior in the laboratory rats; on the other hand, as discussed, we have also learned that delivered in certain contexts, running can become deadly. These lab rat stories suggest that behavior can be both a medicine of sorts and a poison. In the final section of this chapter, we'll explore the therapeutic effects of the behaviors observed in the laboratory rodents.

Behaviorceuticals

I admit it. I made up this word. But why not *behaviorceuticals*? We have pharmaceuticals and, more recently, neutraceuticals. Both of these fields identify drugs or nutrients that enhance the health of individuals or prevent or treat disease. After considering the therapeutic effects of running, there is no doubt that this behavior has a healthy effect on the brain and, although the research is in its infancy, offers considerable promise as a form of treatment or preventative medicine for several neural diseases. If we take these data seriously, which I certainly believe we should, we may be given a prescription for running, or another behavioral response, by our general practitioner or neurologist in the future.

After conducting rat research providing convincing evidence of the brain's ability to rewire itself, Michael Merzenich, at the University of California at San Francisco, became so convinced that behavioral training could be just as therapeutic as drugs or surgery that he began several cutting-edge businesses to research, develop, patent, market, and monitor behavioral therapies for the brain. Merzenich has developed behavioral training strategies for various learning disabilities as well as more severe mental illnesses such as schizophrenia. One of his companies,

Posit Science, has developed engaging interactive software programs. These software programs are described as noninvasive tools to enhance brain health by stimulating the brain's natural plasticity in individuals of all ages. "It's medicine," Merzenich declared in an interview about the therapeutic potential of neuroplasticity. "The more we understand it, the more we understand how to control plasticity to drive targeted changes that drive specific benefits on a more sophisticated and complete level than is generally achieved with drugs." Accordingly, Merzenich and his colleagues are currently running his computer-based training programs through U.S. Food and Drug Administration (FDA) clinical trials just as a promising new drug would be treated before it is deemed safe for human populations.

Let's consider the potential value of behavioral therapies in the case of depression. Although the effectiveness of antidepressants such as fluoxetine (Prozac) remain controversial, research suggests that these drugs (which momentarily increase the amount of the neurotransmitter serotonin in the gaps between neurons) promote neurogenesis, or the growth of new cells. If so, neurogenesis may be an underlying mechanism eventually produced after the administration of antidepressants. Knowing that it takes about four weeks for a newly generated cell to develop into a mature neuron, the four-week lag between taking antidepressant drugs and starting to notice that the depression symptoms are lifting begins to make sense.

In a fascinating study, National Institute on Aging neuroscientist Henriette van Praag and her colleagues pitted antidepressants against running to determine the most effective treatment for depression, at least according to the rodents. She used the common antidepressant Prozac as well as a more recently developed antidepressant, duloxetine (Cymbalta), a drug that increases the amount of norepinephrine as well as serotonin. Rats were given appropriate doses of Prozac, Cymbalta, or running. The production of neural stem cells was increased in the running group at a rate of about 200 percent higher than the other drug and control groups. When the percentage of new cells that matured into neurons was evaluated, the Prozac and running groups produced higher percentages than the control and Cymbalta groups. Focusing on anxiety behavior, the researchers assessed the various groups in an open field so that the amount of time spent in the center, a response associated with lower anxiety levels, could be determined. It was interesting that both Cymbalta and Prozac decreased the amount of time in the center compared to the running group and the control group that received no treatment.

These rat data clearly indicate that running had the most potent effect on cellular proliferation in the brain and that the antidepressant drugs *increased* anxiety in the rats placed in the open field. Although many questions remain as scientists strive to identify the most effective therapies for the pervasive and debilitating depressive disorders, the support for a clear advantage of the most popular antidepressant drugs is waning while there is building excitement among researchers for the therapeutic advantages of running, exercise, and behavioral training. Perhaps there is some merit to the term behaviorceuticals. As is the case with psychoactive pharmaceuticals, behavior certainly has an impact on the brain's chemical and physiological responses. As I expressed in a recent book focusing specifically on the most effective treatments for depression, *Lifting Depression*, our society's willingness to remain faithful to these pharmaceutical treatment approaches is troubling in the face of more and more data supporting less invasive therapeutic strategies.

I hope the laboratory mice and rats have enlightened your thoughts about the best ways to change your brain in positive ways. If nothing else, your brain should be somewhat enriched after enduring all the technical histological terms I've thrown your way in this chapter. After considering the rich literature on neurogenesis and neuroplasticity, one would be hard-pressed to find a single behavior with more positive neural effects than running. In laboratory rodents, both voluntary and forced running increases neurogenesis, along with a host of additional brain boosts. Running increases the blood supply to the brain, decreases the death of existing cells, increases levels of brain fertilizers, generates more complex structures on the dendrites of the brain's neurons . . . the list goes on and on. Yet, with all these positives, the rat data suggest that emotional context is also important, as suggested by the potential fatal effects of the activity–stress paradigm.

I chose the title "Winning the Rat Race" for this chapter due to its inherent double meaning. Typical races are about running, and we have certainly discussed the neurobiological effects of this behavior. The term *rat race* is also linked to

> One would be hard-pressed to find a single behavior with more positive neural effects than running.

notions of lifestyle and purpose and meaningfulness in one's life. Thoughts of living in a rat race are rarely associated with positive thoughts. Just as the rats placed in the activity–stress paradigm worked diligently to improve their conditions by using the wheel to transport them to another place rich with resources, humans can be viewed as using their career treadmills as a form of lifestyle migration to put them in a different place. The success of these career treadmills is sometimes speculative. At times the rewards get confusing to our brains positioned in our contemporary environments. Even if a career led to a higher salary and a nicer house with a staff attending to all our needs, one's brain may not interpret this as "a better place." Considering the toxicity of the activity–stress paradigm, it is clear that our races need to lead to resources valued by our brains, that they actually take us to a better place. Suddenly, winning the rat race is a task that is more complicated and rewarding than any of us ever thought.

As I was collecting information for this chapter, trying to get my hands on everything I could related to running and races and brains and mental health, I serendipitously stumbled on a trailer for a yet-to-be-released documentary about the education system. Although the film had nothing to do with rats, the theme of the documentary, focusing on the harried, tiresome, endless schedules American children face in their schools today, struck a chord. On top of eight hours of school, kids spend hours in sports activities and other lessons, followed by hours of homework. Extended hours of training are important for learning, but do these schedules produce optical learning performance and meaningful achievement? The reason the movie trailer caught my attention was the film's title, *Race to*

Nowhere, a title that makes me think of our children in running wheels going nowhere and, along the way to nowhere, developing physiological and neurological signs of stress. The film's director, Vicki Abeles, a mother of three children, explores the rationale and potential outcomes of the schedules she says "we have all bought into." Thinking back to relevant material we've discussed in this book, the value of physical play (not necessarily organized sports practice) on the development of the brain (discussed in Chapter 6) comes to mind. I was struck by one statement made by a child in Abeles's film: "I can't remember the last time I had a chance to go in the backyard and just run around."

As I often tell my students, the emergence of this exciting research on neuroplasticity comes with a new responsibility and accountability for what we do with our brains. In this chapter we've learned that we need to use our brains to move our bodies around. The responses accompanying the simple act of running or learning new movements enhance our neural networks. In addition, we need to make sure we aren't in a race to nowhere—that our chosen responses lead to brain enhancement and emotional resilience. Now that the laboratory rodents have revealed this valuable information, they have passed the baton to us in this relay race of life; now it is our turn: *Ready, set, go!*

||||||||||||||||||||||||||||||||||||||||||||||||||||||||||||||||||||

CHAPTER 11

Weapons of Mass Destruction

In this chapter I'm going to present one of the most fascinating stories I've discovered over the past several decades of chronicling rat behavior: the story of how some highly trained African rats are being used to detect land mines, undoubtedly one of humans' most inhumane forms of destruction. In short, Good Samaritan rats are helping humans clean up a mess they've created by planting these vicious weapons throughout certain African territories—areas where children play and regrettably have lost life and limbs—all in the name of protecting one's territory.

But first we'll look at territoriality, aggression, and defeat—ours and rodents'. Not to be completely outdone by humans, we'll learn that the rats' famous attack bite is one of their most threatening weapons of mass destruction. And they have one more defense tactic in their toolbox, a tactic that doesn't involve physical weapons or wounds. Read on to find out more about psychological warfare . . . rodent-style.

Patrolling the Borders

In the late 1940s behavioral scientist John Calhoun wondered how rats might respond to living in overcrowded conditions, and he persuaded his Towson, Maryland, neighbor to let him build a rat enclosure on his property to answer this question. Calhoun later referred to his quarter-acre rat enclosure as Rat City and watched as rats freely bred in their new home. Intrigued by the negative effects the crowding seemed to have on rats and mice, Calhoun continued his studies at the National Institute of Mental Health in the 1950s by building room-size pens of freely living rats or mice, rooms he eventually referred to as Rodent Universities.

Calhoun described his findings in these experiments in a landmark *Scientific American* article in 1962 titled "Population Density and Social Pathology." Readers were fascinated to hear all the negative effects Calhoun had observed in his crowded rodent pens, including altered patterns of sexual behavior, maternal neglect, cannibalization of offspring, and heightened male aggression as groups of males appeared to patrol their crowded borders. On all accounts, it seemed that these rodent communities had disintegrated under the pressures associated with overcrowding.

The media loved these findings and spread the message about crowded cities leading to the demise of human morality and health. Negative images of rats going around in gangs attacking females and young males and socially withdrawn animals not able to cope with social stress became popular icons for the potential hazards of city living.

I would argue that, although Calhoun's original study was interesting, it was more a study of chronic stress than crowding per se. Chronic stress conditions, such as prolonged competition for resources and constant uncertainty about social status, associated with either city or country living, can lead to a breakdown in adaptive functions—an observation that few species can escape. As the borders became less defined in the overcrowded rat rooms, relevant and distinct stimuli usually responsible for regulating social functions faded into ambiguity, leading to atypical social responses. The restrictive environment also likely prompted a change in reproductive strategies to enhance the survival of fewer individuals. As we learned in Chapter 9, rats quickly adjust their reproductive strategies to maximize the survival of their offspring.

Another example of rat crowding can be found in the Karni Mata Temple, constructed in India in the early 1900s as a tribute to rats that are believed to be reincarnated ancestors. According to a *National Geographic* article, this rat residence has to be the closest thing to rat heaven on earth. Approximately twenty thousand rats live in this temple, resulting in masses of brown fur running across the marble floors. In this unique temple, where dozens of rats can be seen lapping up sweet milk in communal bowls, human visitors consider the opportunity to taste food or water sampled by the rats to be a blessing. Another special blessing is seeing a white rat—thought to be manifestations of Karni Mata herself. Although to my knowledge no systematic studies have been conducted on the temple rats, there is no evidence of the detrimental behaviors observed by Calhoun. By all accounts, the animals are thriving, increasing their numbers with no evidence of disease.

Despite Calhoun's observations, the occurrence of pathologically aggressive rats is not a typical observation in either laboratory or natural rodent conditions. On the contrary, rats usually resort to aggression and violence only as a last resort. Providing evidence of their social propensities, rats are often observed to be sleeping in huddles, a big pile of rats bunched together. At times the rat on the top will wake up, roll off the top, and then make its way into the bottom of the pile.

When Push Comes to Shove

Despite the pervasiveness of affiliative social interactions, certain situations evoke threat responses and subsequent fights. It is difficult to find a more aggressive animal than a mama rat defending her young, and many researchers have identified key underlying factors related to maternal aggression. However, most of the aggression research has been conducted on males. S. A. Barnett described the interaction between a strange male when placed in another male's cage or territory. After a phase of introductory whiffs of one another, one response typically seen, especially in nonaggressive males, is a behavior aptly named *crawling under* in which the resident rat crawls underneath the strange male's belly. Another variation is the *walking over* response, sometimes accentuated with a bit of urine, the ultimate put-down. In other situations a male may engage in social grooming or nudge the neck area with his nose, a behavior

Barnett described as *nosing*. These behaviors resemble several of the behaviors observed in young rat play responses and appear to be a form of information gathering preceding an actual aggressive encounter.

In some cases, the information gathered during the initial social contact prompts an attack. This is typically due to a robust sexually active male entering another male's territory, prompting the resident male to attack the newcomer to maintain his resources. Alternatively, the presence of a male that isn't sexually mature rarely provokes aggression from the resident male. As described by behavioral neuroscientists Klaus Miczek and Sietse F. de Boer, the males announce an attack with brief, high-pitched vocalizations. As the resident rat becomes increasingly aroused, researchers often observe teeth chattering. So, if you have extraordinary hearing abilities and you ever find yourself approaching a screaming, teeth-chattering rat, take that as a sign to get the heck out of there!

Once the announcement has been made, the resident rat typically exerts his control over the situation, or at least that's how it appears, by grooming the neck region of the intruder rat. Slow-motion video analysis of this response indicates that the resident essentially holds the neck skin between his sharp teeth, prompting a freezing response in his competitor. The similarity between this scene and the classic scenes from movies with the aggressor holding a gun to the victim's head is all too clear. If the intruder rat moves he will get a nasty bite, often accompanied by a swift kick from the resident for good measure.

If events progress, an attack bite, described by Miczek and de Boer as the sine qua non element of the rat fight, will be directed toward the back of the neck. For added style points, an attack jump reminiscent of a *Matrix*-inspired move may precede the attack bite. After all four legs leave the ground the resident lands on the intruder's back, prompting the intruder to send submissive messages by rapidly rolling on its back and exposing its vulnerable belly. At this point a wrestling-style pin is achieved by the resident. By presenting its underside, the rat is hiding the back of the neck, a favorite biting area. This harrowing experience typically consists of five to ten postures and encounters in some orderly sequence—sniffing pursuit, threat posture, attack bite, subsequent threat postures, pin—rarely exceeding about thirty seconds. Game over.

The lesson from our discussion of these aggressive encounters of rats is clear: Their weapon of mass destruction is their front teeth. And if you haven't had an opportunity to open a rat's mouth and peek in, they have some formidable teeth. The real threats are the four incisors; the upper two are about four millimeters long and the lower incisors are about seven millimeters long. Ouch! Another weapon, however, is much less physical; in fact, it could be argued to be all in their little rodent heads, all the more interesting since this psychological weapon can be deadly.

Psychological Warfare

When I was in graduate school conducting research on activity stress and gastric ulcers in rats, my good friend Kim Levy Huhman focused her research attention on another rodent, the Syrian hamster. I know what you're thinking, awww . . . how cute.

Hamsters have a secret.

Indeed, they are high on the cuteness scale for rodents, but hamsters have a secret. They are *extremely* aggressive. Kim had not chosen this species due to their attractiveness; on the contrary, it was their predisposition toward aggressive encounters that sealed the deal. Unlike the social tendencies of rats, Syrian hamsters are known to be solitary mammals that work to defend their territory from all animals. If a poor unsuspecting intruder is dropped in a resident hamster's cage, it's a sure bet that he's going to receive a physical beating. But the fighting response isn't nearly as interesting as the hamster form of mental anguish that follows this initial defeat encounter.

To be specific, if an intruder hamster is placed in the home cage of another hamster (resident hamster), the resident will defend his territory by beating up the unwelcomed visitor. The interesting part of this story is the severe impact this loss has on the loser. After being placed in its own home cage and given ample time to recover, the loser hamster acts atypically when a smaller hamster is placed in the cage. The loser fails to defend his territory like any self-respecting resident male hamster should do. It's as if the hamster, after a single defeat by another male hamster, learned that it was a loser; thus this phenomenon is known as *conditioned defeat*. This lesson seems to stick with loser hamsters for as long as thirty-three days.

Huhman's team discerned that social defeat is accompanied by somewhat of a tsunami of neurochemical effects, including increased stress neurochemicals such as adrenocorticotropic hormone (ACTH), ß-endorphin, and glucocorticoids and decreased levels of testosterone. Merely placing the defeated hamster in the presence of the resident male, even when there is a physical barrier to prevent any harmful contact, triggers the defeat neurochemical profile. Hence, although this stress begins with a physical encounter, it persists without the contact, providing evidence of what these researchers refer to as *psychological stress*. Negative health effects emerge subsequent to the defeat experience, effects likely due to drastic alterations of neurochemicals involved in stress and learning responses. Defeated hamsters develop high blood pressure, increased heart rate, and suppressed immune function.

Because this effect is so robust in male hamsters—virtually 100 percent of the intruder males exhibit conditioned defeat— it is interesting that female hamsters, also aggressive animals, fail to consistently show this behavior. In one study, only about 28 percent of the females exhibited conditioned defeat and, when they did show it, there were no long-term neurochemical effects.

This research reminds us of the power of a single traumatic experience, how an encounter lasting mere seconds can leave a long-term neurochemical footprint, drastically altering an animal's behavior.

Bully Rats

The conditioned defeat research also brings another topic to mind: social bullying. Research suggests that human adolescents who are not only victims of social bullying but bullies themselves are more likely to suffer from symptoms of depression and even suicide. In humans, females don't appear to be more resilient in bullying situations, as observed in the hamsters. In fact, evidence suggests a heightened vulnerability in adolescent girls. The intensity and pervasiveness of the emotional effects of social threats can be appreciated by the clues that have been presented about social defeat in the Georgia hamsters. These emotions run deep and define future social encounters in these animals. There is, however, a glimmer of hope for us humans: Unlike hamsters, we are social animals; friendships have been shown to mitigate some of the sting associated with social stress.

What about the effects of this psychological stress in other rodent models? Eberhard Fuchs and his colleagues at the Laboratory of Clinical Neurobiology in Gottingen, Germany, have provided evidence that social stress can be life threatening, even in social rats. In his lab's rat protocol, the residents are large male rats (weighing about three hundred grams) that have been housed with females to heighten their aggressive and territorial tendencies. Later, the female is removed from the rat's cage and a smaller adult male (about two hundred grams) is placed in the cage. You can guess what happened after that. After about a minute, the big resident rat will beat up the smaller visiting rat. For the remainder of an hour, the losing rats are placed in a small wire mesh cage and kept in the resident rat's larger cage. The cage creates a physical barrier that protects the intruder rat, preventing the likelihood of physical harm. This scenario was repeated for five weeks, at which time the data suggest that this is very stressful for the rats, leading to behavioral signs of depression and neural restructuring, such as decreased motivation to consume a sugary water treat and a smaller hippocampus (brain area involved in emotions and learning), respectively. This line of research may cause us to reconsider jobs that force workers to reside in tiny cubicles within shouting distance of hostile bosses!

Thus research conducted on both laboratory and wild rodents provides clear evidence that, in the right conditions, rats are indeed aggressive. Different from many human encounters, however, the rodent encounters are often swift and result in limited physical injuries. In some conditions, especially contrived laboratory manipulations, the mere presence of a more dominant rodent peer can wreak havoc on the health of the less dominant animal. This psychological warfare, accompanied by a deluge of stress hormones, often leads to compromised mental functions, such as the emergence of depression.

As I conclude this section on rodent aggression, it is important to reiterate that, unlike reptiles, rodent aggression isn't an automatic reflexive response triggered by the mere presence of another animal. Conditions have to be just right for a fight to be triggered, and even then, the fight follows sufficient posturing and exploring to confirm the necessity of the battle. It is interesting that two large breeding male rats, the equivalent of the Hulk Hogans of the species, will not fight if placed in a new environment at the same time. No established territory, no reason to fight. Thus humans may learn a thing or two from rats about reserving aggressive responses for when they are absolutely necessary for protecting

No established territory, no reason to fight.

valuable resources. And there are other lessons we can learn from the rodents about fighting battles. As promised, in the next section we'll discuss how rats have donated their sophisticated weapon-detection abilities to reverse some of the potential damage humans inflict on one another; enter the HeroRATs!

The HeroRATs

The first time I read about HeroRATs, I thought I had stumbled on a piece of fiction. A Belgian social philanthropist, Bart Weetjens, who loved his rodents as a boy, learned that he could successfully train African giant pouch rats to detect TNT in land mines. He subsequently created a company, APOPO, to train and license rats to locate land mines in Africa. APOPO stands for Anti-Persoonsmijen Ontmijnende Productonwikkeling, Flemish for antipersonnel mines demining product development. Weetjens was attracted to the African rats because they were easy to train, had a long life span (up to eight years), and were extremely sociable. Using Pavlovian techniques known by general psychology students across the world, he developed a protocol to painstakingly train rats in his lab in Tanzania, established in collaboration with the Sokoine University of Agriculture. When rats approached the TNT odor, they would hear a clicker that had previously been associated with a yummy treat, often a piece of banana. In anticipation of the reward, the rats would scratch the surface surrounding the highly valued TNT cup. This behavior became incorporated into the conditioning sequence. Smell TNT, scratch surface, get reward. Behavioral aficionados will recognize another famous style of conditioning in this protocol, Skinnerian operant conditioning. Once the associations between TNT and food rewards are established, the animal has to respond with the gentle digging motion before receiving the treat. The classical conditioning element of this training involves the association between TNT vapors and the clicker/treat. The operant conditioning element involves the association between the digging response and the presentation of the yummy reward.

After about a year of training, the HeroRAT special agents are ready for some fieldwork. In a *Frontline* segment, the rats are shown being harnessed so they can be guided along a patch of land that has been partitioned for assessment. And Weetjens's team didn't forget the sunscreen for the rats' ears; the rodents are nocturnal and not used to being out in the sunlight.

The rat's job in the field training is to do exactly what it was taught in the lab, search for TNT (located in buried tea canisters in this phase of training) and scratch the dirt when identified. And, most important for the rat, sit and wait for the valuable reward. The tendency of these rats to store their food underground in their natural habitats facilitates the ease at which these behaviors can be trained. The APOPO team spoke of individual differences among the rats, who were given names such as Utomi, Ararat, and even Obama. In the interview, there were high hopes for Ararat who exhibited perfect performance in his field tests. It is important to point out that no rats have lost their lives in their bomb-sniffing field trips; although they are large for rats, they are too small to detonate the land mines. Dogs could also be trained for this task but their weight is a concern as well as their emotional bonds to a single trainer.

Rats are more likely to work for many different people as long as the training conditions are similar.

The HeroRAT graduates have identified many land mines in Mozambique that were subsequently carefully diffused by their human coworkers. This heroic rodent effort has allowed APOPO to give land, including playgrounds, back to the residents of Africa. The HeroRATs have been accredited by the Mine Action Standards Pact as a preferred technology for clearing land mines. To date, the HeroRATs have returned 1,312,000 square meters of treasured land back to its African inhabitants.

Also impressive, HeroRATs are helping humans fight the war against immunological threats. APOPO has commenced training the rats to detect tuberculosis (TB) in human saliva. Their ability to smell *Myobacterium tuberculosis* is a much more efficient laboratory assay than human laboratory assistants' use of visual screening techniques, even when their detection capabilities are facilitated with a microscope. Whereas it takes humans one day to run twenty saliva samples, the rats can complete up to two thousand per day. "Rats with Noses Beat Humans with Microscopes" would be an interesting headline!

In 2009, Weetjens received the Skoll Award for Social Entrepreneurship. According to Weetjens, his goal, with the help of the rats, of course, is to develop affordable, appropriate technology to empower local communities to regain access to the valuable land that has been marred with dangerous land mines. The idea of rats contributing to international award-winning social entrepreneurship endeavors is certainly an unexpected surprise for the Lab Rat Chronicles!

After learning about Weetjens's impressive rodent training

program (and watching an entertaining video tribute to them on YouTube), I couldn't help but think about another research area that is often on my mind. Once trained, these rats are employed by APOPO for several years until they are too old to work. That's right, the rats have a lifetime full of training and working for their rewards, a wonderful example of effort-driven rewards. I contacted Weetjens, and we quickly began designing a study to compare the working rats to similar-aged rats that hadn't been exposed to the training and work schedules of their bomb-sniffing counterparts—a real-life investigation of the working rat versus the trust fund rat research my students and I have been exploring in the lab. We have yet to secure funding for these international rat research ideas, but we'll continue to persist. I look forward to meeting the Tanzanian working rats some day in the future.

CHAPTER 12

||||||||||||||||||||||||||||||||||||||||||||||||||||||||||||||||||||||||||||||||||||

Keeping It Real: The Most Important Lesson of All

Throughout this book I have conveyed my enthusiasm for the lessons we can learn from laboratory rats. These accolades were prompted by the interesting and adaptive responses exhibited by rats when confronted with a variety of tests. As impressed as I've been with the behavior of laboratory rats, however, my thoughts keep wandering to wild rats. I currently don't have enough material for *The Wild Rat Chronicles* as so much more research has been conducted in the laboratory than in the field. But there's a spattering of research and observations of wild rats that provide additional information about our mammalian ancestral roots. Although I've included a few wild rat highlights along the way, in this chapter we'll focus more exclusively on wild rats, contemplating the cost of domestication for both rats and humans. This analysis raises some big questions about the impact of industrialization, modern conveniences, and readily available, brain-altering pharmaceuticals and their effects on the never-ending evolution of our brains and our abilities.

You may recall the wild rat described in Chapter 1 that broke into my office and lab several years ago—a visit that planted the seeds for this book. In preparation to write this chapter, I decided to return the favor by visiting the wild rats' territory, a guided tour of the rat-infested back alleys of Baltimore—what an adventure!

Survivor: Rodent Edition

Most of us have asked ourselves the question, What if I were stranded on a deserted island—no human contact, no readily available food sources, no premanufactured housing—could I survive even for a day?

When a male Norway rat wandered into a trap baited with chocolate on a forested New Zealand island in 2004, he had no idea how his life was about to change. Spy tracking devices, search dogs, poison, and a team of scientists would soon become the context of his life, and his triumph would give him a superhero status in a children's book. But I've skipped the details—let's get back to that piece of chocolate, a treat that distracts humans and rats alike.

After following his nose to a chocolate-baited live trap, a group of University of Auckland scientists anesthetized their rodent hostage and proceeded with a battery of tests. Not unlike science-fiction movies in which alien captors log data from the earth inhabitants, the scientists recorded the rat's sex, body length, and weight. They also retrieved a small tissue sample from the tip of his tail for DNA genetic fingerprinting so the animal could be positively identified when subsequently captured. The final task involved fitting the rat with a radio collar so he could be easily tracked across the island.

Forty-eight hours after that fateful bite of chocolate, the rat awoke on the beaches of Motuhoropapa, a twenty-four-acre island he had never visited in his short life. The reason for this rat espionage drama? The scientists were concerned about the rats' repeated attempts to successfully invade the islands and wanted to know more about how a solitary rat would fare in the many survivor challenges they had in store for him. To answer their questions, a little spying was necessary.

For the first twenty-six days, the movement of the rat was followed closely until it was established that the rat was consistently traversing a home range of an area measuring less than half an acre. Similar to the procedures used to track a suspect in crime dramas such as *CSI*, the team of scientists carefully tracked the rat hour by hour to learn his pattern of activity and then carefully determined strategic areas to place various traps with goodies such as peanut butter and chocolate. According to the scientists, these traps represented "the best-practice devices in rodent detection." The plan was to check the traps nightly until the rat suspect was apprehended, a task that wasn't thought to take more than a couple of days. If ever there was an unfair advantage, the humans, with their huge brains and fancy tracking and trapping devices, appeared to be more than a formidable match for the 250-gram rat that had been plucked from his densely populated rat community and dropped on a deserted island of which he had no prior knowledge. Another strike against him was the fact that he had been previously trapped, indicating that this was an animal that would let down his guard for a piece of chocolate, a sign that he wasn't the most clever rat in the pack.

After eight nights of searching, the scientists doubled their trapping efforts, leaving more traps baited with other enticing treats such as salmon and salami. Weeks passed and new traps were set out, but the rat eluded the traps night after night, week after week. Rodent-detection dogs, trained fox and border terrier crosses, were brought in. After ten long weeks, the scientists realized that they had lost the rat's radio signal; there was no sign that the rat was on the island. What did this mean? The rat wasn't located in any of the traps that had been strategically placed along the frequented areas of the island. Even if the rat had died, the collar could still be tracked. What could explain this mysterious disappearing act?

The answer came about two weeks later from the inhabitants of the neighboring island of Otata. The residents reported strange bird activity. The team of scientists and their high-tech gear descended on this island and collected some rat fecal samples. After running the necessary tests, the scientists determined that the DNA matched the rat they had released on Motuhoropapa. The answer to this CSI mystery was becoming increasingly clear. Although he wasn't voted off the island, for some reason the adventurous rat took the plunge into open water and swam 438 meters across unchartered waters to an unknown destination. This was the first documented case of a rat swimming hundreds of meters across open water. The next mystery for these scientists was, Why? Why leave the comforts of an island with food and shelter to dive into unknown waters? The best answer they had was that he was lonely, or in science lingo, they conveyed that a lack of conspecifics sent him on his journey.

As these answers emerged, the team of scientists still faced a lack of ability to track the resilient and clever animal. For the next four weeks, they would enhance their efforts once again, introducing new traps and new alluring treats. They brought in the dog squad one more time and paid close attention to an area that seemed to be rich in rat odor. They doubled their efforts once again by placing traps strategically along that area. Eighteen weeks into the trapping adventure, the rat was successfully captured; fresh penguin meat was his weakness. Quite an exotic palate for a rodent! The scientists summed up this experience in their *Nature* paper with this anticlimactic statement: "Our findings confirm that eliminating a single rat can be disproportionately difficult." Yes, you might say that. This rat's adventure will live on—not only in the scientific literature but in the children's book *The Amazing Adventures of Razza the Rat* written by New Zealand author Witi Ihimaera.

The adaptive responses of the New Zealand rat provide compelling evidence that the wild rats are indeed survivors. Being able to navigate and locate resources in a novel environment and elude significant threats to survival suggest that wild rats have what it takes to persist, survive, and thrive in challenging environments. As we look to the domesticated laboratory rats, however, I wonder how much of their wild rat ancestral ghosts have been retained. After all, the rats I use in my lab represent hundreds of generations of animals bred over the past half a century that have never

Wild rats are indeed survivors.

known any environment that even remotely resembles the wild. Do they have what it takes to survive a wilderness challenge? Oxford University animal behaviorist Manuel Berdoy was also curious about that question and decided to put it to the test.

The Lab Rats Earn Their Wilderness Merit Badge

To determine just how much of their ancestral past the laboratory rats had retained, Berdoy and his colleagues selected a barnyard that was fenced off from most predators and released both albino and hooded laboratory rats. They just took the laboratory cages to the site and removed the lids— bon voyage to the rats! As the rats cautiously emerged from their familiar cages, they sniffed the air before boldly venturing out. They had never felt sun or wind or seen the grass or sky; this was certainly a brave new world for these rats that had previously lived such a sheltered life. The rats' behavior in this novel environment was videotaped and was ultimately released as a documentary that my laboratory students view each year.

Although the laboratory rats have no doubt been influenced by generations of domestication, they remain the sophisticated product of millions of years of evolutionary pressures (as is the case for humans as well). This field experiment would reveal just how deep the evolutionary roots run: Would these rats be able to find food, water, and shelter in this new environment? This is a question I often ask myself as a researcher. Just how close to the wild rats are these docile rats in my laboratory? Is their behavior truly representative of the wild rats? I couldn't wait to see how these Oxford University rats would fare in the real world.

After exiting the laboratory cages the first order of business appeared to be locating available shelter. They cautiously and awkwardly climbed on various structures and located hiding places in the readily available bales of hay. Locating a water source was next on their to-do list; once they located a water source, one rat was shown jumping in. Others sat along the sidelines drinking. The individual differences with a response as basic as drinking were remarkable. Forced to lick a tube emerging from a water bottle their whole lives, they now had options. Some leaned over and lapped up the water like a cat or dog. Others used a more sophisticated method of scooping the water and drinking out of their paws as if they were sipping tea from a cup—very Beatrix Potter–esque. After a few short hours, the rats' movement patterns started to look different. Their cautious walking responses were transformed into a hopping gait as they appeared to be bouncing across the terrain. And there was the digging, lots of digging as if they were making up for lost time. The digging eventually led to the creation of burrows in the hay bales—a form of luxury apartments for the rats.

What about food? There were no manufactured nutritionally balanced rat pellets in the new environment. They encountered eggs, berries, and roots, among other things. Some finds were edible, some weren't. A few rats managed to crack the mystery of getting to the egg yolks. Many of the rats were immediately interested in the berries, but were cautious. The hooded rats, with their keener vision, oriented toward the ripe berries and began with small tastes. Other, less bold rats followed and sniffed and observed, then proceeded to eat the berries. The albino rats took longer to catch on to the idea that the dark berries were less bitter but eventually learned to select the ripe berries. Other objects such as roots were sampled but later determined to be unsuitable for food. The rats' omnivore preferences provided ample opportunities for securing food in this new environment.

Environmental challenges also presented themselves. A summer rainstorm was definitely a new experience. It is safe to say that most labs do a good job avoiding this scenario! The rats responded by seeking shelter in the hay. Then the scientists stepped up the challenges by introducing a cat. Uh-oh. These rats had never been exposed to a cat. Would they know that this animal was a predator? Would a few rats be killed before the rats would learn to avoid cats? The suspense was killing me! At the first sign of the cat—likely its odor—every rat bolted into their burrows, remaining hidden for about twenty-four hours.

After a few days, visible signs of an established rat colony were apparent. Elaborate burrows and nesting sites were observed throughout the hay bales. Transportation runs, almost exactly the width of the rats, were created in the tall grass to facilitate their movement among several important destinations in the new environment. The night videos showed rats quickly transporting themselves from destination to destination like a mini-subway system. And, generally, there was a lot of activity observed during the dark hours. This work schedule offered protection against predators that relied on vision to locate them, such as birds of prey and cats. The rats became noticeably more agile as they navigated their environment, but they also increased their vigilance when venturing out as they seemed to be becoming increasingly aware of the many threats in their new world.

Stable social structures also began to emerge. Occasional aggressive encounters were observed but were rare. Males were obviously interested in the females. A single female rat was shown darting and running across the environment with a string of hopeful males following after her. The leader of the pack was the one who copulated with her first. After about three weeks, pups were born. At this time there was clear evidence that the white albino and the black and white hooded rats had no sexual discriminatory practices as they interbred freely.

So, after two hundred generations or so of captivity, the laboratory rats appeared to have retained their ancestral roots. After just a few hours, they appeared to be transformed into feral animals. After six months, the rats continued to thrive.

> **The laboratory rats appeared to have retained their ancestral roots.**

The documentary I show my students provides reassurance that the laboratory rat remains a representative model of *Rattus norvegicus*. The narrator ends the film by suggesting that, although we had taken the rat out of its wild environment, we hadn't taken the wild out of the rat. In fact, in a recent email, Manuel Berdoy told me that he had studied lots of species—from baboons in Africa to penguins in Antarctica—and he had the most respect for the wild rats he has studied in his own backyard around Oxford University.

There was one behavior in the released rats, however, that continued to differentiate the lab from the wild rats. The lab rats remained less fearful of novel objects and were quicker to approach them than their wild counterparts. Although this characteristic could be viewed as bold and adaptive, as it is often classified in the laboratory, it likely doesn't serve the rat well in the real world as paranoia in wild rats is a sign of a healthy grasp of reality. That's right—although extreme paranoia can be debilitating, a healthy dose of skepticism often differentiates the surviving rats from the rats killed by traps, poisons, or other interesting but ultimately dangerous animals.

These wild rat tales reignited my desire to observe the rats in their natural habitat. I value the findings from laboratory studies but certainly realize that they will be enhanced if scientists identify and incorporate rats' natural tendencies into their studies. Although videos and scientific reports are great resources to learn more about wild rats, I needed to be closer to these animals in their actual environments, even if it was just a brief encounter.

My Visit to the Rat Alleys

If you're interested in wild city rats, Greg Glass, a professor in the Department of Microbiology and Immunology at Johns Hopkins University, is your guy. Because one of my former students, Sabra Klein, is in the same department and collaborates with Glass, I had heard stories about the Baltimore rats and his unorthodox research practices for several years. Whereas most scientists order animals that arrive in neatly packed boxes, this is certainly not the case for Glass's research team. They patrol the most likely alleys for rodent habitation (often the most dangerous areas in Baltimore) to select the most ideal habitats to set traps at sunset. The traps are then retrieved at sunrise and the ensnared city rats are subsequently brought into the lab where, reminiscent of a forensic autopsy on your favorite criminal investigation drama, the animals are examined from nose to tail. After the basic height and weight stats are recorded, their fur is combed to identify any parasites, wounds, or scars. Blood samples are taken so a panel of viral and bacteria screens can be used to determine health risks for humans. Further, both blood and fecal samples are used for hormone tests to determine levels of stress and sex hormones. The bodies are then dissected so the researchers can examine various organs. The thymus gland and spleen provide information about immune functioning, and the adrenal gland (which secretes stress and resiliency hormones) reveals information about stress responsiveness. Thus a lot can be learned by reading the many clues provided by these wild rat bodies.

Recently, Glass graciously agreed to take an afternoon and evening to show me the Baltimore rats' world. Curiously, there was a list of directions for me: I was instructed to wear running shoes and to remove all jewelry when going out in the field. I was also told that it was a good idea to hide an extra form of identification and twenty dollars in my shoe just in case we were mugged and needed to take a taxi back to Hopkins. Wow, I never had to wear running shoes in the lab for fear of actually having to run for my life; entering the world of the wild urban rats was going to be quite an adventure. I couldn't wait!

The contrast between the beautiful high-tech building that

houses Glass's and Klein's labs and offices in the Johns Hopkins Bloomberg School of Public Health and the surrounding residential alleys containing the rats was striking. Although Glass explained in our afternoon tour that the conditions had improved dramatically, evidence of rodent inhabitants was still apparent. The alleys were located behind residential areas made up of row houses with very small backyards containing concrete parking areas and small garden patches. As we walked through the alleys, Glass pointed out that the water running through the middle of the alley, spillage from a pipe, was an essential environmental feature for rats, a must-have before they commit to their new real estate. Then there was the garbage stacked where rats could easily have access. As we looked closer to the yard areas, golf ball–size holes were everywhere—formed in the crumbling concrete, in the aged wooden fences, and in the dirt around the residents' gardens. I learned that the rats were happy to move completely underground when the housing opportunities became restricted aboveground. Many of these small backyards were merely caps on complex rat burrows beneath. In fact, one of the alleys had actually collapsed several years ago when a garbage truck drove through because the rats had undermined the concrete drive with their extensive tunnels.

During our afternoon tour there was no evidence of live rats (they're nocturnal), but a few dead ones in the middle of the alley were apparent. Their gray fur looked different from the albino and hooded rats I work with in my lab but everything else was the same; *Rattus norvegicus* with a different fur color. Glass explained that many of the rats were caught by cats but only up until about two months of age, after that they were too big for the cats to take on. But there were many other threats facing these contemporary city rats: diseases, aggressive encounters with other rats and animals, traps, and poison, for example. In short, the rats' real world, at least for these urban dwellers, is tough, so tough that Glass's research indicates that a significant portion of the rats don't survive past two months of age. Drastically different from my laboratory rats, these alley rats needed every competitive advantage they could muster if they wanted to survive to reproductive maturity and beyond in the real world.

After the hot August afternoon stroll, we climbed into Glass's truck for a field trip to other alleys that had been tagged prime rodent territory. As we parked and cautiously walked through the alleys at sunset, I began to see the live rats quickly scurrying across the alleys. We noticed the alley cats vigilantly watching certain patches of grass, patches that, on closer observation, had rats running through them. I felt my adrenaline surge and the team laughed as I reflexively jumped in excitement at my first glimpse of the adaptive creatures making their way in the demanding city. We visited several alley sites throughout the evening, and each one gave me new insight about the real world of contemporary urban rats. The last alley we visited, Druid Hill, was apparently a little riskier than the others so we drove through, stopping on occasion to wait for the inevitable rat running across the alley. I couldn't help but feel that I was on some unique urban safari—having to stay in the vehicle not because of the danger presented by the rats but, in this case, the unpredictable behavior of the human inhabitants.

My urban rat education continued through dinner as my colleagues described some notable physical features of the wild rats. Their adrenal glands, important for their function relating to the release of stress hormones, were roughly the size of a blueberry, about ten times larger than the shriveled-up adrenal glands in laboratory rats. This was also the case for the spleen, known for its role in immune functions. It was remarkably larger in the wild rats. These observations provide compelling evidence that the wild rat is indeed encountering many more threats, both physical and psychological, than are the laboratory rats. And there was something else, this time relating to diet and physical fitness. Just cutting through the abdominal muscles is a very different task in a wild rat than in a lab rat. "They're ripped," Klein declared as she went on to describe the muscles sported by the wild rats. They also lacked the white fatty tissue spread throughout their abdominal cavity, a characteristic feature of our sedentary, well-fed lab rats. Their large intestines provided additional interesting clues. In the many lab rats I have dissected through the years I often note how the large intestine looks like a string of linked sausages because there is an abundance of fecal pellets, or poop, backed up in the intestines waiting to be excreted. (Yes, this was our dinner conversation.) The intestines were clean in the wild rats, suggesting that there was a closer match between energy consumed and energy expended; they were more likely to eat what they needed for their active lifestyle and not to overindulge.

Hmmm, I couldn't help but think if my physical characteristics would have any resemblance to my genetically matched human ancestors who occupied the natural world before our technological interventions. Am I closer to a lab rat or a wild rat? Could I survive if dropped on a deserted island with a team of a more intelligent species following my every move and plotting to use the best technology to trap me? Am I sufficiently cautious when faced with new, potentially dangerous stimuli such as food or other objects? Have I kept enough of my evolutionary wits about me to ensure survival in unpredictable circumstances that I may face in the future? Would I be the first of the cast to be voted off the island in *Survivor*?

Since my alley rat safari, I have been increasingly aware of the natural resilience gained by our ancestors' evolutionary journey. Although I genuinely appreciate the protection modern society offers against the threats faced by the wild rats, including predator and immune threats, I can't help but feel that there is a limit to just how cushy we can make our lives without compromising our survival potential. How has our increasing reliance on modern conveniences and mind-altering drugs altered our brains—brains that haven't changed in any substantive way since our ancestors stepped out of the natural environment into high-rises? How important is it to retain at least some natural neural and behavioral tendencies?

Green Brains: Nature's Inconvenient Truth

Former Vice President Al Gore's 2006 documentary *An Inconvenient Truth* brought the facts of a changing climate to the attention of the earth's citizens. The question of interest was the potential impact industrialization and our increasing dependence on carbon-based fuels have had on the earth's infrastructure. A similar question can be asked about the lasting impact of our industrialized lifestyles on the infrastructure of our brains. In my book *Lifting Depression*, I argue that the increasingly sedentary lifestyles that characterize the past century are linked to increased rates of depression in our society. At least the rat data suggest this is the case. The nonworking trust fund rats described in Chapter 3 have lower physiological measures of resilience and exhibit less persistence when new challenges are presented. Thus the lab rats suggest that our increased dependence on conveniences associated with our modern lifestyles may have a significant impact on our brain's infrastructure, or at least its functional output.

University of Wisconsin biological paleoanthropologist John Hawks's work suggests that the impact of changing lifestyles, for good or bad, began changing our brains long before the appearance of modern conveniences. It is surprising that an analysis of skulls from our human ancestors dating back five thousand years ago suggest that the human brain has shrunk about 10 percent. "As to why is it shrinking, perhaps in big societies, as opposed to hunter-gatherer lifestyles, we can rely on other people for more things, can specialize our behavior to a greater extent, and maybe not need our brains as much," Hawks said in a recent interview. To support this view, domesticated animals have long been observed to have smaller brains than their wild counterparts.

As a student of the brain, I find the notion of shrinking brains fascinating. I've been in this field long enough to know that a shrinking brain may not necessarily be a bad thing. After all, the brain is an energy hog requiring a disproportionate share of fuel. If, through evolution, the brain has found a way to streamline its energy requirements while maintaining its functional capabilities, that would be an advantageous modification. At this point, however, we don't know this is the case and it's very likely that there are functional costs to our neuronal downsizing.

Like the city rats I observed, we're caught in a type of environmental and cultural limbo, somewhere between a domestic and wild animal. Although the urban rats are closer to the wild category, many of them take advantage of human-generated food and shelter resources, just as we do. This seems adaptive in that they have ready-made resources, but are there any costs? The life spans of wild rats living in the country are longer than those of city rats; in fact, country rats have frequently been observed to live up to eighteen months.

Catherine Franssen, a postdoctoral assistant, and I are attempting to address this cultural confusion by designing studies in which we are constructing natural enriched environments consisting of all natural elements such as dirt, sticks, moss, and rocks. These animals are compared to animals in more traditional enriched environments, consisting of plastic and metal objects. For example, in the climbing category in these environments, the natural animals may have a stick, whereas the artificial animals have a plastic ladder. For the bedding category, the natural animals may have moss, but the artificial animals are given cotton balls. The animals placed in the mixed environment are exposed to half-artificial and half-natural stimuli. The items are changed a couple of times a week so that novelty is still a component of the enrichment. Several students have worked on this pioneering line of research; they like to refer to the animals housed in the natural environments as the "country rats," those housed in the artificial environment as the "city rats," and the mixed-environment animals as the "suburban rats." Our data are still rolling in but the early results suggest that the type of enriched environment has little impact on learning or cognitive ability but it does affect more basic emotional responding.

For example, the country rats had lower stress hormones after a rigorous swim versus their city and suburban counterparts. So, it may be that, whereas a city rat is

The country rat may have an emotional advantage.

just as likely to be as smart as a country rat, the country rat may have an emotional advantage.

The issue of green brains is brought into sharp focus when we consider the astounding role pharmaceuticals have come to play in defining our mental health over the past several decades. Pharmaceutical interventions no doubt offer promise and hope for many challenges faced by our brains, but for some conditions, such as depression, the impact is less clear. In *Anatomy of an Epidemic,* journalist Robert Whitaker chronicles our society's relationship with antidepressants and antipsychotics used to treat mental illnesses such as depression and schizophrenia. Even though we spent twenty-five billion dollars on these drugs in 2007, the rates of individuals diagnosed with mental illness and receiving long-term government assistance (such as Social Security Disability Insurance) have skyrocketed. Since the FDA approval of the antidepressant Prozac in 1987, the number of mentally disabled individuals has doubled and since 1955 has increased by sixfold.

According to the lab rats, there is no clear evidence that drugs such as Prozac, known as selective serotonin reuptake inhibitors (SSRIs), are effective for depression-like symptoms; further, there is even less evidence suggesting that long-term use is either safe or effective. For example, research suggests that rats placed on

SSRI treatment results in the animals having reduced levels of fear conditioning; this sounds like a welcoming drug effect, but remember what we learned from the wild rats—a certain amount of fear is essential for real-world survival. In addition to our modern conveniences, are psychoactive drugs accelerating the domestication of our brains? With at least 10 percent of our population taking antidepressants, are we becoming increasingly docile and less fearful of stimuli that pose real threats to us, just as we have observed in the domesticated lab rats? The messages being sent by the rodent studies should cause us to approach the decision to influence our brain's natural fuel production with extreme caution. The massive increases in the use of antidepressant drugs, up 75 percent since 1996, bring visions of a chemical cage being dropped on our society, restricting the brain's ability to maximize its functional potential should we ever find ourselves in more demanding circumstances.

Take-Home Point: Maintaining Our Evolutionary Wits

Ultimately, the lessons emerging from the rodents remind us of the basics, of nature's simple truths that have survived the test of time. According to the informative research conducted by so many scientists across the world, the "Rodent Self-Help Best Practices" include the following observations:

- **Challenging and Complex Environments Build Better Brains.** Rats exposed to enriched environments have larger, healthier brains.

- **Practice May Not Make Perfect, but It Transforms Brains.** Cognitive training builds neural networks and shifts response strategies in relevant and adaptive ways.

- **Relevant and Rewarding Work Provides an Emotional Lift.** Work-like experiences that emphasize relationships between effort and positive consequences build resilience, likely by enhancing a sense of control in stressful situations.

- **Healthy Lifestyles Are Nature's Best Medicine.** There is considerable value in preventative healthcare, starting with healthy diets and lifestyle habits; rats are extremely cautious when eating a newly introduced food.

- **Day-to-Day Coping Strategies Are Essential for Avoiding Stress-Related Diseases.** Being a flexible coper, responding when necessary and holding back when appropriate, is related to healthier emotional and physiological responses.

- **Social Networks Define Neural Networks.** Positive social relationships—among family members, sexual partners, and conspecific friends—alter neurochemistry and neuroanatomy in healthy ways.

- **Cleanliness Promotes Healthy Bodies and Minds.** Good hygiene practices, even in rats, are related to healthy emotional regulation and can facilitate social interactions.

- **Physical Activity Promotes Brain Growth.** Rats allowed to run in activity wheels show evidence of new brain cells and healthy brain landscapes.

- **The Races We Run Should Have Clear Destinations.** Rats housed in activity wheels and placed in artificial feeding situations enhance their running to dangerous levels.

- **When It Comes to Aggression, Less Is Better.** Rats are extremely conservative in confrontational situations, resorting to physical attack only when absolutely necessary.

- **Psychological Stress Is Toxic.** Even when no physical harm is inflicted, rats and other rodents that are placed in situations in which they have little control (for example, being housed in the presence of a dominant conspecific) suffer from extreme stress-related illnesses.

- **Nature's Neurochemicals Represent the Best Pharmacological Treatments.** Natural responses appear to be the most effective and relevant modulators of the brain's chemistry in normally functioning rat brains; therefore, extreme caution should be taken before introducing exogenous mind/brain-altering chemicals, for recreational or medicinal purposes.

- **Family Trees Have Deep Roots and Deserve Our Respect.** Certain elements of the rat ancestor's natural environments are relevant for their contemporary brains; thus certain aspects of our ancestors' environments should be included in modern lifestyles.

Although humans are obviously different from rats in many ways, I have come to believe they can teach us a lot about how to live a healthy, happy, and productive life. And the scientific literature continues to grow. There's no telling where the next rat adventures will take us!

REFERENCES

Chapter 1. Whisker Wisdom Emerges

3   **stowing away on ships:** A. N. Iwaniuk, "Evolution," in *The Behavior of the Laboratory Rat*, edited by B. Kolb and I. Whishaw, 3–14, Oxford: New York, 2005; and J. R. Lindsey and H. J. Baker, "Historical Foundations," in *The Laboratory Rat*, 2nd ed., edited by M. A. Suckow, S. H. Weisbroth, and C. L. Franklin, 1–52, New York: Academic Press, 2006.

3   **rats fell on hard times:** J. R. Lindsey and H. J. Baker, "Historical Foundations," 1–52.

3   **rat fighting:** R. Sullivan, *Rats: Observations on the History and Habitat of the City's Most Unwanted Inhabitants*, New York: Bloomsbury, 2004.

4   **Swiss psychiatrist Adolph Meyer:** J. R. Lindsey and H. J. Baker, "Historical Foundations," 1–52.

4   **85 percent of all biomedical research:** H. Wurbel, C. Burn, and N. Latham, "The Behaviour of Laboratory Mice and Rats," in *The Ethology of Domestic Animals*, 2nd ed., edited by P. Jensen, 217–233, Cambridge: Cabi, 2009.

4   **one scientific publication based on the responses of laboratory rats:** M. Berdoy, *The Laboratory Rat: A Natural History*, produced by Manuel Berdoy in affiliation with Oxford University, National Television System Committee, video, 2002.

5   **smaller and more economical to house, mice:** A. N. Iwaniuk, "Evolution," 3–14.

5   **the rat chose complexity:** I. Q. Whishaw, "The Laboratory Rat, the Pied Piper of Twentieth Century Neuroscience," *Brain Research Bulletin* 50 (1999): 411.

5   **ascendancy from the gutter to a place of nobility:** J. R. Lindsey and H. J. Baker, "Historical Foundations," 1–52, quotation from p. 2.

9   **the effects of a plethora of drugs:** National Heart, Lung and Blood Institute, "NIH Meeting on Rat Model Priorities," May 3, 1999: www.nhlbi.nih.gov/resources/docs/ratmtg.htm.

12   **animals exposed to chronic stress:** D. Fleming and K. G. Lambert, unpublished findings, 2007.

13   **Instead of giving up immediately:** K. G., Lambert, K. Tu, A. Everette, G. Love, I. McNamara, M. Bardi, and C. H. Kinsley, "Explorations of Coping Strategies, Learned Persistence, and Resilience in Long-Evans Rats: Innate vs. Acquired Characteristics," in *Resilience in Children*, edited by B. M. Lester, A. Masten, and B. McEwen, 319–324, *Annals of the New York Academy of Sciences*, vol. 1094, New York: New York Academy of Sciences, 2006.

13   **the behaviors described:** Rat Genome Sequencing Consortium, R. A. Gibbs, et al. "Genome Sequence of the Brown Norway Rat Yields Insights into Mammalian Evolution," *Nature* 428 (2004): 493–521, doi:10.1038/nature02426; and "Scientists Decode Rat Genome," *Voice of America News*, April 1, 2004: www.voanews.com/english/news/a-13-a-2004-04-01-4-1-66345582.html; and H. Wurbel, C. Burn, and N. Latham, "The Behaviour of Laboratory Mice and Rats," 217–233.

14   **enhanced adaptation rate led to enhanced survival rates:** H. R. Pilcher, "Rat Genome Unveiled," *Nature News*, April 1, 2004, doi:10.1038/news040329-11.

14   **the genus *Rattus*, with its approximately fifty-five species:** M. Berdoy and L. C. Drickamer, "Comparative Social Organization and Life History of *Rattus* and *Mus*," in *Rodent Societies: An Ecological and Evolutionary Perspective*, edited by J. Wolff and P. Sherman, 380–392, Chicago: University of Chicago Press, 2007; and "Scientists Decode Rat Genome," source of Collins quotation.

15   **neuroscientist Kent Berridge:** K. C. Berridge, "Measuring Hedonic Impact in Animals and Infants: Microstructure of Affective Taste Reactivity Patterns," *Neuroscience and Biobehavioral Reviews* 24 (2000): 173–198.

15   **Neuroscientist Jaak Panksepp:** J. Panksepp, "Neuroevolutionary Sources of Laughter and Social Joy: Modeling Primal Human Laughter in Laboratory Rats," *Behavioral Brain Research* 182 (2007): 231–244.

16   **How do you know when rats are laughing?** J. Panksepp and J. Burgdorf, "Laughing Rats? Playful Tickling Arouses High-Frequency Ultrasonic Chirping in Young Rats," *American Journal of Play* (winter 2010), 357–372.

17   **the human brain's tendency to distort reality:** S. Aamodt and S. Wang, *Welcome to Your Brain*, New York: Bloomsbury, 2008.

18   **John de Castro's work:** "Dean Offers Eating Tips for Busy Students," *Sam Houston State University News*, November 4, 2009: www.shsu .edu/~pin_www/T@S/2009/eatingtips.html.

18   **biases and inaccurate strategies exist for adaptive reasons:** K. E. Stanovich, *What Intelligence Tests Miss*, New Haven, CT: Yale University Press, 2010.

19   **the rats of New York City:** R. Sullivan, *Rats*, quotation from p. 218.

Chapter 2. Building the Brain Trust

22   *anthropomorphism:* L. Daston and G. Mitman, eds., *Thinking with Animals*, New York: Columbia University Press, 2005.

24   **young rats in a cage with lots of toys:** E. L. Bennett, M. C. Diamond, D. Krech, and M. R. Rosenzweig, "Chemical and Anatomical Plasticity of the Brain," *Science* 164 (1964): 610–619; and M. C. Diamond, *Enriching Heredity: Impact of the Environment on Brain Development*, New York: Free Press, 1988; and M. C. Diamond, D. Krech, and M. R. Rosenzweig, "The Effects of an Enriched Environment on the Rat Cerebral Cortex," *Journal of Comparative Neurology* 123 (1964): 111–119; and R. L. Holloway, "Dendritic Branching: Some Preliminary Results of Training and Complexity in Rat Visual Cortex," *Brain Research* 2 (1966): 393–396; and K. Mollgaard, M. C. Diamond, E. L. Bennett, M. R. Rosenzweig, and B. Lindner, "Quantitative Synaptic Changes with Differential Experience in Rat Brain," *International Journal of Neuroscience* 2 (1971): 113–128.

26   **Diamond conveyed how scary it was:** Marian Diamond, personal communication, October 2009.

26   **the brain was fixed:** S. Finger, *Minds Behind the Brain: A History of the Pioneers and Their Discoveries,* New York: Oxford University Press, 2000.

27   **naturalist and children's writer Beatrix Potter:** L. Lear, *Beatrix Potter: A Life in Nature,* New York: St. Martin's Griffin, 2007, quotation from p. 44.

27   **early suggestions of environmental influences:** M. C. Diamond, *Enriching Heredity*; and D. Hebb, *The Organization of Behavior: A Neuropsychological Theory,* New York: Wiley, 1949.

28   **Even more exciting is the observation:** J. E. Black, K. R. Isaacs, B. J. Anderson, A. A. Alcantara, and W. T. Greenough, "Learning Causes Synaptogenesis, Whereas Motor Activity Causes Angiogenesis, in Cerebellar Cortex of Adult Rats," *Proceedings of the National Academy of Sciences U.S.A.* 45 (1990): 307–312.

28   **reduce the symptoms of depression:** K. Lambert, *Lifting Depression: A Neuroscientist's Hands-On Approach to Activating Your Brain's Healing Power,* New York: Basic Books, 2008.

29   **Neuroscientists in India:** J. Veena, B. N. Srikumar, T. R. Raju, and B. S. Shankaranarayana Rao, "Exposure to Enriched Environment Restores the Survival and Differentiation of New Born Cells in the Hippocampus and Ameliorates Depressive Symptoms in Chronically Stressed Rats," *Neuroscience Letters* 455 (2009): 178–182.

29   **rats placed on certain work schedules:** M. Felton and D. O. Lyon, "The Post-Reinforcement Pause," *Journal of the Experimental Analysis of Behavior* 9 (1966): 131–134; and C. B. Ferster and B. F. Skinner, *Schedules of Reinforcement,* New York: Appleton-Century-Crofts, 1957.

30   **study conducted by David Foster and Matthew Wilson:** D. J. Foster and M. A. Wilson, "Reverse Replay of Behavioural Sequences in Hippocampal Place Cells During the Awake State," *Nature* 440 (2006): 680–683.

31   **looking through the microscope is like:** Paul MacLean, personal communication, April 2001.

32   **When interviewed about this study:** V. Limjoco, "Coffee Break Brain," *ScienCentral*, June 23, 2006: www.sciencentral.com/articles/view.php3?type=article&article_id=218392812.

32   **three hours to devote to studying:** K. C. Bloom and T. J. Shuell, "Effects of Massed and Distributed Practice on the Learning and Retention of Second-Language Vocabulary," *Journal of Educational Research* 74 (1981): 245–248.

33   **the coffee part of the coffee break:** C. T. Wentz and S. S. P. Magavi, "Caffeine Alters Proliferation of Neuronal Precursors in the Adult Hippocampus," *Neuropharmacology* 56 (2009): 994–1000.

34   **creative work conducted by Jonathon Crystal and Allison Foote:** J. D. Crystal and A. L. Foote, "Metacognition in Animals: Trends and Challenges," *Comparative Cognition and Behavior Reviews* 4 (2009): 54–55; and A. L. Foote and J. D. Crystal, "Metacognition in the Rat," *Current Biology* 17 (2007): 551–555.

37   ***Enriching Heredity*, Diamond reported:** M. C. Diamond, *Enriching Heredity*.

Chapter 3. Neuroeconomics and Long-Term Investments

39   **video clip of a mother rat:** *The Life of Mammals*, hosted by David Attenborough, BBC, video, 2003.

41   **Neuroscientist Paul Glimcher:** P. Glimcher, *Decisions, Uncertainty, and the Brain*, Cambridge, MA: MIT Press, 2003.

42   **Alison Fleming:** A. Lee, S. Clancy, and A. S. Fleming, "Mother Rats Bar-Press for Pups: Effects of Lesions of the MPOA and Limbic Sites on Maternal Behavior and Operant Responding for Pup-Reinforcement," *Behavioural Brain Research* 100 (1999): 15–31.

43   **Joan Morrell and her colleagues:** M. P. Wansaw, M. Pereira, and J. P. Morrell, "Characterization of Maternal Motivation in the Lactating Rat: Contrasts Between Early and Late Postpartum Responses," *Hormones and Behavior* 54 (2008): 294–301.

49   **worker rats spent about 60 percent more time:** K. G. Lambert, K. Tu, A. Everette, G. Love, I. McNamara, M. Bardi, and C. H. Kinsley,

"Explorations of Coping Strategies, Learned Persistence, and Resilience in Long-Evans Rats: Innate vs. Acquired Characteristics," 319–324.

49   **the effects of neuropeptide Y:** C. Morgan, S. Wang, S. M. Southwick, A. Rasmusson, G. Hazlett, R. I. Hauger, and D. S. Charney, "Plasma Neuropeptide Y Concentrations in Humans Exposed to Military Survival Training," *Biological Psychiatry* 47 (2000): 902–909; and T. J. Sajdyk, P. L. Johnson, R. J. Leitermann, S. D. Fitz, A. Dietrich, M. Morin, D. R. Gehlert, J. H. Urban, and A. Shekhar, "Neuropeptide Y in the Amygdala Induces Long-Term Resilience to Stress-Induced Reductions in Social Responses but Not Hypothalamic-Adrenal-Pituitary Axis Activity or Hyperthermia," *Journal of Neuroscience* 28 (2008): 893–903.

49   **effort-driven reward therapy enhances resilience:** A. Rhone, M. Bardi, C. L. Franssen, E. S. Shea, J. E. Hampton, M. M. Hyer, J. Huber, and K. G. Lambert, "Recipe for Resilience: Explorations of Coping Strategies and Effort-Driven Reward Training in Male Long-Evans Rats (*Rattus norvegicus*)," poster presented at the International Behavioral Neuroscience Society annual meeting, Sardinia, Italy, 2010.

50   **neuroscientists James Olds and Peter Milner:** J. Olds and P. Milner, "Positive Reinforcement Produced by Electrical Stimulation of Septal Area and Other Regions of Rat Brain," *Journal of Comparative and Physiological Psychology* 47 (1954): 419–427.

51   **neuroscientist John Salamone:** J. D. Salamone, "The Behavioral Neurochemistry of Motivation: Methodological and Conceptual Issues in Studies of the Dynamic Activity of Nucleus Accumbens Dopamine," *Journal of Neuroscience Methods* 64 (1996): 137–149; and J. D. Salamone, M. Correa, A. M. Farrar, E. J. Nenes, and M. Pardo, "Dopamine, Behavioral Economics, and Effort," *Frontiers in Behavioral Neuroscience* 3 (2009), doi:10.3389/neuo.08.013.2009.

53   **The forebrain has been implicated in effort-related decision making:** J. D. Salamone, M. Correa, A. Farrar, and S. M. Mingote, "Effort-Related Functions of Nucleus Accumbens Dopamine and Associated Forebrain Circuits," *Psychopharmacology* 191 (2007): 461–482.

53   **Tibor Scitovsky:** T. Scitovsky, *The Joyless Economy: The Psychology of Human Satisfaction*, New York: Oxford University Press, 1992, quotation from p. 71.

54   **Justin Joffe designed a neighborhood:** J. M. Joffe, R. A. Rawson, and J. A. Mutlick, "Control of Their Environment Reduces Emotionality in Rats," *Science* 180 (1973): 1383–1384.

54   **Neuroscientist Kent Berridge:** K. C. Berridge, "Pleasures of the Brain," *Brain and Cognition* 52 (2003): 106–128.

55   **neuroscientist Dennis Choi:** "New Field Works to Understand the Complexities of Decision-Making," *Emory Magazine*, March 20, 2009: http://shared.web.emory.edu/emory/news/releases/2009/03/new-field-in-complexities-of-decision-making.html.

55   **financial journalist Michael Lewis:** M. Lewis, *The Big Short: Inside the Doomsday Machine*, New York: Norton, 2010.

56   **he collected functional MRI (fMRI) readings:** J. B. Engelmann, C. M. Capra, C. Noussair, and G. S. Berns, "Expert Financial Advice Neurobiologically 'Offloads' Financial Decision-Making Under Risk," *PLoS ONE* 4, no. 3 (2009): e4957, doi:10.1371/journal/pone.0004957.

Chapter 4. Universal Healthcare

59   **although we spend more than twice:** T. Farley and D. A. Cohen, *Prescriptions for a Healthy Nation: A New Approach to Improving Our Lives by Fixing Our Everyday World*, Boston: Beacon Press, 2005; and World Health Organization, "Top Ten Causes of Death," Fact Sheet no. 310, November 2008: www.who.int/mediacentre/factsheets/fs310.pdf.

61   *omnivore's paradox:* P. Rozin, "The Selection of Foods by Rats, Humans and Other Animals," *Advances in the Study of Behavior* 6 (1976): 21–76.

61   **a new food substance:** S. A. Barnett, *The Rat: A Study in Behavior*, Chicago: University of Chicago Press, 1975.

62   **experimental psychologist William James:** Ibid, quotation from p. 34.

62 **John Garcia's discovery:** K. B. Freeman and A. L. Riley, "The Origins of Conditioned Taste Aversion Learning: A Historical Analysis," in *Conditioned Taste Aversion: Behavioral and Neural Processes*, edited by S. Reilly and T. R. Schachtman, 9–36, New York: Oxford University Press, 2009.

66 **rats become addicted to sugar:** N. M. Avena, P. Rada, and B. G. Hoebel, "Evidence for Sugar Addiction: Behavioral and Neurochemical Effects of Intermittent, Excessive Sugar Intake," *Neuroscience and Biobehavioral Reviews* 32 (2008): 20–39.

67 **evidence of food addiction:** P. M. Johnson and P. J. Kenny, "Dopamine D_2 Receptors in Addiction-Like Reward Dysfunction and Compulsive Eating in Obese Rats," *Nature Neuroscience* 13 (2010): 635–641.

68 **a lawsuit was filed against McDonald's:** S. Bernstein, "McDonald's Faces Lawsuit over Happy Meals," *Los Angeles Times*, June 23, 2010: www.latimes.com/business/la-fi-0623-happy-meals-20100623,0,4821950.story.

69 **Sarah Leibowitz's lab:** G. Z. Chang, V. Gaysinskaya, O. Karatayev, and S. F. Leibowitz, "Maternal High-Fat Diet and Fetal Programming: Increased Proliferation of Hypothalamic Peptide-Producing Neurons That Increase Risk for Overeating and Obesity," *Journal of Neuroscience* 28 (2008): 12107–12119.

69 **pregnant women in Nairobi:** Marian Diamond, personal communication, November 2009; and M. C. Diamond, "Successful Aging of the Healthy Brain," paper presented at the Conference of the American Society on Aging and the National Council on the Aging, New Orleans, March 10, 2001: www.newhorizons.org/neuro/diamond_aging.htm.

71 **costs associated with obesity are outrageous:** E. A. Finkelstein, I. C. Fiebelkorn, and G. Wang, "National Medical Spending Attributable to Overweight and Obesity: How Much, and Who's Paying?" *Health Affairs* W3 (2003): 219–226.

71 **the National Institute on Drug Abuse (NIDA) recently reported:** NIDA, "Drug Abuse and Addiction: One of America's Most

Challenging Public Health Problems," National Institutes of Health, June 2005: http://archives.drugabuse.gov/about/welcome/about drugabuse/magnitude.

71 *Perscription for a Healthy Nation:* T. Farley and D. A. Cohen, *Prescription for a Healthy Nation*, quotation from p. 133.

71 **a whopping 20 percent of adults:** *Journal of American Medical Association*, state-specific prevalence and trends in adult cigarette smoking—United States (1998–2007): 302 and (2009): 250–251.

72 **a preference test:** J. Wolffgramm and A. Heyne, "From Controlled Drug Intake to Loss of Control: The Irreversible Development of Drug Addiction," *Behavioural Brain Research* 70 (1995): 77–94.

72 **injections of drugs in certain environments:** M. T. Bardo and R. A. Bevins, "Conditioned Place Preference: What Does It Add to Our Preclinical Understanding of Drug Reward?" *Psychopharmacology* 153 (2000): 31–43.

73 **Bruce Alexander:** B. K. Alexander, B. L. Beyerstein, P. F. Hadaway, and R. B. Coambs, "Effect of Early and Later Colony Housing on Oral Ingestion of Morphine in Rats," *Pharmacology, Biochemistry, and Behavior* 15 (1981): 571–576; and J. C. Brenes and J. Fornaguera, "Effects of Environmental Enrichment and Social Isolation on Sucrose Consumption and Preference: Associations with Depressive-Like Behavior and Ventral Striatum Dopamine, *Neuroscience Letters* 436 (2008): 278–282; and C. R. De Carvalho, P. Pandolfo, F. A. Pamplono, and R. N. Takahashi, "Environmental Enrichment Reduces the Impact of Novelty and Motivational Properties of Ethanol in Spontaneously Hypertensive Rats," *Behavioral Brain Research* 208 (2010): 231–236; and S. Peele and B. K. Alexander, "Theories of Addiction," in *The Meaning of Addiction*, edited by S. Peele, 47–72, San Francisco: Jossey-Bass, 1998; and M. Pilkington, "Lab Habits: Do Depressed Lab Rats Dictate International Drug Policy?" *Guardian*, June 2005: www.guardian.co.uk/science/2005/jun/02/farout; and S. Raz and B. D. Berger, "Social Isolation Increases Morphine Intake: Behavioral and Psychopharmacological Aspects," *Behavioral Pharmacology* 21 (2010): 39–46.

74   **higher addiction rates have been reported in prisoners:** A. H. Malcolm, "Explosive Drug Use Creating New Underworld Prisons," *New York Times*, December 30, 1989: www.nytimes.com/1989/12/30/us/explosive-drug-use-creating-new-underworld-prisons.html.

76   **the importance of lifestyle factors:** L. N. Robins, D. H. Davis, and D. W. Goodwin, "Drug Use by U.S. Army Enlisted Men in Vietnam: A Follow-Up on Their Return Home," *American Journal of Epidemiology* 99 (1974): 235–249.

77   **social isolation, Alexander argues, has contributed:** B. K. Alexander, *The Roots of Addiction in Free Market Society*, 6–7, Vancouver, BC: Canadian Centre for Policy Alternatives, 2001.

77   **George Koob and his colleagues:** O. George, S. Chozland, M. R. Azar, P. Cottone, E. P. Zorrilla, L. H. Parsons, L. O'Dell, H. N. Richardson, and G. F. Koob, "CRF-CRF1 System Activation Mediates Withdrawal-Induced Increases in Nicotine Self-Administration in Nicotine-Dependent Rats," *Proceedings of the National Academy of Sciences U.S.A.* 104 (2007): 17198–17203.

78   **psychologist Robert Ader:** R. Ader, "Conditioned Immunomodulation: Research Needs and Directions," *Brain, Behavior and Immunity* 17 (2003): S51–S57.

80   **seminal article was published in the journal *Psychosomatic Medicine*:** R. Ader and N. Cohen, "Behaviorally Conditioned Immunosuppression," *Psychosomatic Medicine* 37 (1975): 333–340.

81   **removal of germs from one's environment:** A. P. Devalapalli, A. Lesher, K. Shieh, J. S. Solowk, M. L. Everett, A. S. Edala, P. Whitt, R. R. Long, N. Newton, and W. Parker, "Increased Levels of IgE and Autoreactive, Polyreactive IgG in Wild Rodents: Implications for the Hygiene Hypothesis," *Scandinavian Journal of Immunology* 64 (2006): 125–136.

Chapter 5. Enhancing Emotional Resilience

83   **Craig Kinsley and I wrote a textbook:** K. G. Lambert and C. H. Kinsley, *Clinical Neuroscience: The Neurobiological Foundations of Mental Health*, New York: Worth, 2005.

84   **costs associated with stress-related illnesses:** B. McEwen and E. N. Lasley, *The End of Stress as We Know It*, Washington, DC: Henry Press, 2002.

85   **your dentist's speaking skills may be more important:** A. Ploghaus, C. Narain, C. F. Beckmann, S. Clare, S. Bantick, R. Wise, P. M. Matthews, J. N. Rawlins, and I. Tracey, "Exacerbation of Pain by Anxiety Is Associated with Activity in a Hippocampal Network," *Journal of Neuroscience* 21 (2001): 9896–9903.

86   **Selye was excited:** R. M. Sapolsky, *Why Zebras Don't Get Ulcers*, New York: Henry Holt, 1993.

88   **bypassed naming this new concept:** C. L. Cooper, *Theories of Organizational Stress*, New York: Oxford University Press, 1998.

88   **Selye misunderstood the English-language descriptions:** W. K. Smith, "The Stress Analogy," *Schizophrenia Bulletin* 13 (1987): 215–220.

89   **Selye's contributions to this field:** B. McEwen and E. N. Lasley, *The End of Stress as We Know It*.

89   **Peter Sterling, a neuroscientist:** P. Sterling, "Principles of Allostasis: Optimal Design, Predictive Regulation, Pathophysiology and Rational Therapeutics," in *Allostasis, Homeostasis, and the Costs of Adaptation*, edited by J. Schulkin, 17–64, New York: Cambridge University Press, 2004.

90   *allostatic load:* B. S. McEwen, "Allostasis and Allostatic Load: Implications for Neuropsychopharmacology," *Perspectives* 22 (2000): 108–124.

91   **stress hormone corticosterone into the rats:** C. S. Woolley, E. Gould, and B. S. McEwen, "Exposure to Excess Glucocorticoids Alters Dendritic Morphology of Adult Hippocampal Pyramidal Neurons," *Brain Research* 53 (1990): 225–231.

92   **determine if actual psychological stress:** Y. Watanabe, E. Gould, and B. S. McEwen, "Stress Induces Atrophy of Apical Dendrites of Hippocampal CA3 Pyramidal Neurons," *Brain Research* 588 (1992): 341–345.

92   **similar shrinking effects:** K. G. Lambert, K. M. Gerecke, P. S. Quadros, E. Doudera, A. M. Jasnow, and C. H. Kinsley, "Activity-Stress Increases Density of GRAP-Immunoreactive Astrocytes in the Rat Hippocampus," *Stress* 34 (2000): 275–285.

93   **nerve cells are actually dying:** C. D. Conrad, "What Is the Functional Significance of Chronic Stress-Induced CA3 Dendritic Retraction within the Hippocampus?" *Behavioral and Cognitive Neuroscience Reviews* 5 (2006): 41–60; and M. H. Kole, B. Czeh, and E. Fuchs, "Homeostatic Maintenance in Excitability of Tree Shrew Hippocampal CA3 Pyramidal Neurons After Chronic Stress," *Hippocampus* 14 (2004): 742–751.

93   **conducted with Martha McClintock:** S. A. Cavigelli and M. K. McClintock, "Fear of Novelty in Infant Rats Predicts Adult Corticosterone Dynamics and an Early Death," *Proceedings of the National Academy of Sciences U.S.A.* 100 (2003): 16131–16136.

95   **females who spontaneously developed:** S. A. Cavigelli, J. R. Yee, and M. K. McClintock, "Infant Temperament Predicts Life Span in Female Rats That Develop Spontaneous Tumors," *Hormones and Behavior* 50 (2006): 454–462.

95   **Richard James was working:** M. D. Baum, "It's Slinky, It's Slinky," February 11, 2001: http://archives.cnn.com/2001/US/02/10/slinky.story; and D. Hevesi, "Betty James, Who Named the Slinky Toy, Is Dead at 90," *New York Times*, November 24, 2008.

97   **procedure known as the *back test*:** J. M. Koolhaas, S. F. de Boer, C. M. Coppens, and B. Buwalda, "Neuroendocrinology of Coping Styles: Towards Understanding the Biology of Individual Variation," *Frontiers in Neuroendocrinology*, 2010, doi:10.1016/j.yfrne.2010.04.001; and M. J. C. Hessing, A. M. Hagelso, J. A. van Beek, P. R. Wiepkema, W. G. P. Schouten, and R. Krukow, "Individual Behavioural Characteristics in Pigs," *Applied Animal Behavior Science* 37 (1993): 285–295.

97   **wishy-washy group was viewed as "doubtful":** E. Van Erp-Van der Kooij, H. H. Kuijpers, J. W. Schrama, E. D. Ekkel, and M. J. M. Tielen, "Individual Behavioural Characteristics in Pigs and Their

Impact on Production," *Applied Animal Behavior Science* 66 (2000): 171–185.

98 **undergraduate student Kelly Tu:** K. G. Lambert, K. Tu, A. Everette, G. Love, I. McNamara, M. Bardi, and C. H. Kinsley, "Explorations of Coping Strategies, Learned Persistence, and Resilience in Long-Evans Rats: Innate vs. Acquired Characteristics," 319–324.

100 **Darby Fleming Hawley, spearheaded that effort:** D. F. Hawley, M. Mardi, A. M. Everette, T. J. Higgins, K. M. Tu, C. H. Kinsley, and K. G. Lambert, "Neurobiological Constituents of Active, Passive, and Variable Coping Strategies in Rats: Integration of Regional Brain Neuropeptide Y Levels and Cardiovascular Responses," *Stress* 13 (2010): 172–183.

100 **viewed the floating response:** H. D. Schmidt and R. S. Duman, "The Role of Neurotrophic Factors in Adult Hippocampal Neurogenesis, Antidepressant Treatments and Animal Models of Depressive-Like Behavior," *Behavioural Pharmacology* 18 (2007): 391–418.

102 **infusion of this substance into certain areas:** T. J. Sajdyk, P. L. Johnson, R. J. Leitermann, S. D. Fitz, A. Dietrich, M. Morin, D. R. Gehlert, J. H. Urban, and A. Shekhar, "Neuropeptide Y in the Amygdala Induces Long-Term Resilience to Stress-Induced Reductions in Social Responses but Not Hypothalamic-Adrenal-Pituitary Axis Activity or Hyperthermia," 925–935.

102 **special forces soldiers:** C. A. Morgan, S. Wang, S. M. Southwick, A. Rasmusson, G. Hazlett, R. L. Hauger, and D. S. Charney, "Plasma Neuropeptide-Y Concentrations in Humans Exposed to Military Survival Training," *Biological Psychiatry* 47 (2000): 902–909.

103 **an effort-driven reward group:** A. Rhone, M. Bardi, C. L. Franssen, E. S. Shea, J. E. Hampton, M. M. Hyer, J. Huber, and K. G. Lambert, "Recipe for Resilience: Explorations of Coping Strategies and Effort-Driven Reward Training in Male Long-Evans Rats (*Rattus norvegicus*)."

107 **neuroscientist Trevor Robbins:** J. W. Dalley, R. N. Cardinal, and T. W. Robbins, "Prefrontal Executive and Cognitive Functions in

Rodents: Neural and Neurochemical Substrates," *Neuroscience and Biobehavioral Reviews* 28 (2004): 771–784.

109   **psychologist James Gross:** J. J. Gross, "Emotion Regulation: Affective, Cognitive, and Social Consequences," *Psychophysiology* 39 (2002): 281–291.

Chapter 6. The Value of Social Diplomacy

111   **a type of play fighting:** S. M. Pellis and V. Pellis, "Play and Fighting," in *The Behavior of the Laboratory Rat*, edited by B. Kolb and I. Whishaw, 2005, New York: Oxford University Press, 298–306.

114   **Jaak Panksepp, currently at the Washington State University:** N. S. Gordon, S. Kollack-Walker, H. Akil, and J. Panksepp, "Expression of c-Fos Gene Activation During Rough and Tumble Play in Juvenile Rats," *Brain Research Bulletin* 57 (2002): 651–659, quotation from p. 658.

115   **increased release of neurotrophic growth factors:** N. S. Gordon, S. Burke, H. Akil, S. J. Watson, and J. Panksepp, "Socially-Induced Brain 'Fertilization': Play Promotes Brain Derived Neurotrophic Factor Transcription in the Amygdala and Dorsolateral Frontal Cortex in Juvenile Rats," *Neuroscience Letters* 341 (2003): 17–20.

115   **Play-deprived rats:** M. Potegal and D. Einon, "Aggressive Behavioral in Adult Rats Deprived of Playfighting Experiences as Juveniles," *Developmental Psychobiology* 22 (1989): 159–172.

116   **abundant play will facilitate:** J. Panksepp, "Can Play Diminish ADHD and Facilitate the Construction of the Social Brain?" *Journal of Canadian Academy of Child and Adolescent Psychiatry* 12 (2007): 57–66.

116   **the number of elementary schools reporting:** K. R. Ginsburg, "The Importance of Play in Promoting Healthy Child Development and Maintaining Strong Parent-Child Bonds," *American Academy of Pediatrics* 119 (2007): 182–191.

116   **A recent *USA Today* article:** E. Bazar, "Not It! More Schools Ban Games at Recess," *USA Today* (2006): www.usatoday.com/news/health/2006-06-26-recess-bans_x.htm.

117   research suggests that there is a strong connection: A. Diamond, "Close Interrelation of Motor Development and Cognitive Development and of the Cerebellum and Prefrontal Cortex," *Child Development* 71 (2000): 41–56.

117   children are currently taking psychostimulants: W. W. Beatty, K. B. Costello, and S. L. Berry, "Suppression of Play Fighting by Amphetamine: Effects of Catecholamine Antagonists, Agonists, and Synthesis Inhibitors," *Pharmacology Biochemistry and Behavior* 29 (1984): 747–755; and W. W. Beatty, A. M. Dodge, L. J. Dodge, K. Whike, and J. Panksepp, "Psychomotor Stimulants, Social Deprivation and Play in Juvenile Rats," *Pharmacology Biochmemistry and Behavior* 16 (1982): 417–422.

118   literature about play and classroom behavior: C. Blake, C. L. Franssen, J. E. Hampton, and K. G. Lambert, "Rough-and-Tumble Play Improves Performance in the Attention Set Shifting Task," poster presented at the International Behavioral Neuroscience Society annual meeting in Sardinia, Italy, 2010.

121   *Last Child in the Woods*: R. Louv, *Last Child in the Woods: Saving Our Children from Nature-Deficit Disorder*, Chapel Hill, NC: Algonquin, 2008.

123   first case of potential social reciprocity: C. Rutte and M. Taborsky, "Generalized Reciprocity in Rats," *Public Library of Science (PLoS)* 5, no. e196 (2007): 1421–1425.

123   a coin in a public telephone: A. M. Isen, "Positive Affect, Cognitive Processes and Social Behavior," *Advances in Experimental Social Psychology* 20 (1987): 203–258.

124   the contagion of threatening painful responses: D. J. Langford, S. E. Crager, Z. Shehzad, S. G. Smith, S.G. Sotocinal, J. S. Levenstadt, M. L. Chanda, D. J. Levitin, and J. S. Mogil, "Social Modulation of Pain as Evidence for Empathy in Mice," *Science* 312 (2006): 1967–1970.

125   communicative value of facial expressions: D. J. Langford, A. L. Bailey, M. L. Chanda, S. E. Clarke, T. E. Drummond, S. Echols, S. Glick, J. Ingrao, T. Klassen-Ross, M. L. LaCroix-Fralish, L.

Matsumiya, R. E. Sorge, S. G. Sotocinal, J. M. Tabaka, D. Wong, A. M. J. M. van den Maagdenberg, M. D. Ferrari, K. D. Craig, and J. S. Mogil, "Coding of Facial Expressions of Pain in the Laboratory Mouse," *Nature Methods*, 2010, doi:10.1038/NMETH.1455.

127   **role of social contact in wound healing:** H. E. Ashley, I. McNamara, M. Morgan, S. L. Klein, E. R. Glasper, A. C. DeVries, A. F. Conway, K. Tu, C. H. Kinsley, and K. G. Lambert, "Sex and Social Housing Influence Wound Healing and Anxiogenic Responses in Chronically Stressed *Peromyscus californicus*," unpublished manuscript.

128   **DeVries and Glasper conducted:** E. R. Glasper and A. C. DeVries, "Social Structure Influences Effects of Pair-Housing on Wound Healing," *Brain, Behavior and Immunity* 19 (2005): 61–68.

129   **DeVries has written extensively:** A. C. Courtney, T. K. S. Craft, E. R. Glasper, G. N. Neigh, and J. K. Alexander, "Social Influences on Stress Responses and Health," *Psychoneuroendocrinology* 32 (2007): 587–603.

129   **laboratory of Yasushi Kiyokawa:** Y. Kiyokawa, Y. Takeuchi, and Y. Mori, "Two Types of Social Buffering Differentially Mitigate Conditioned Fear Responses," *European Journal of Neuroscience* 26 (2007): 3606–3613.

130   **Martha McClintock's laboratory:** J. R. Yee, S. A. Cavigelli, B. Delgado, and M. K. McClintock, "Reciprocal Affiliation Among Adolescent Rats During a Mild Group Stressor Predicts Mammary Tumors and Lifespan," *Psychosomatic Medicine* 70 (2008): 1050–1059.

131   **McClintock's team confirmed:** G. L. Hermes, B. Delgado, M. Tretiakova, S. A. Cavigelli, T. Krausz, S. D. Conzen, and M. K. McClintock, "Social Isolation Dysregulates Endocrine and Behavioral Stress While Increasing Malignant Burden of Spontaneous Mammary Tumors," *Proceedings of the National Academy of Science U.S.A.* 52 (2009): 22393–22398.

132   **no new news for humans:** J. S. House, K. R. Landis, and D. Umberson, "Social Relationships and Health," *Science* 241 (1988): 540–545.

Chapter 7. The Perils of Bad Hair Days

135 **rat grooming is important:** A. Denmark, D. Tien, K. Wong, A. Chung, J. Cachat, J. Goodspeed, C. Grimes, M. Elegante, C. Suciu, S. Elkhayat, B. Bartels, A. Jackson, M. Rosenberg, K. M. Chung, H. Badani, F. Kadri, S. Roy, J. Tan, S. Gaikwad, A. Stewart, I. Zapolsky, T. Gilder, and A. V. Kalueff, "The Effects of Chronic Social Defeat Stress on Mouse Self-Grooming Behavior and Its Patterning," *Behavioural Brain Research* 208 (2010): 553–559.

136 **typical five-second grooming sequence:** J. W. Aldridge, "Grooming," in *The Behavior of the Laboratory Rat*, edited by I. Q. Whishaw and B. Kolb, 141–149, New York: Oxford University Press, 2005.

137 **reputable rodent grooming researchers:** K. C. Berridge, "Comparative Fine Structure of Action: Rules of Form and Sequence in the Grooming Patterns of Six Rodent Species," *Behaviour* 113 (1990): 21–56.

137 **seventy thousand hair salons:** Hoovers, "Hair Salons: Industry Description": www.hoovers.com/industry/hair-salons/1213-1.html.

138 **the basal ganglia and closely related areas:** J. W. Aldridge, "Grooming," 141–149; and K. C. Berridge and J. C. Fentress, "Disruption of Natural Grooming Chains After Striatopassidal Lesions," *Psychobiology* 15 (1987): 336–342.

142 **Decades of research on this topic:** C. Caldji, B. Tannenbaum, S. Sharma, D. D. Francis, P. M. Plotsky, and M. J. Meaney, "Maternal Care During Infancy Regulates the Development of Neural Systems Mediating the Expression of Behavioral Fearfulness in Adulthood in the Rat," *Proceedings of the National Academy of Sciences U.S.A.* 95, no. 5 (1998): 5335–5340; and D. D. Francis, U. J. Dioriok, D. Liu, and M. J. Meaney, "Nongenomic Transmission Across Generations in Maternal Behavior and Stress Responses in the Rat," *Science* 286 (1999): 1155–1158; and J. Menard, D. Champagne, and M. J. Meaney, "Maternal Care Alters Behavioral and Neural Activity Patterns in the Defensive Burying Paradigm," *Neuroscience* 129 (2004): 297–308.

142   **the offspring of high-licking mothers:** M. Meaney, "Epigenetics and the Biological Definition of Gene X Environmental Interactions," *Child Development* 81 (2010): 41–79.

142   **through *epigenetics*:** Ibid.

143   **specific genetics events accompanying:** L. Buchen, "In Their Nurture," *Nature* 467 (2010): 146–148.

144   **final genetics lessons rats have taught us:** D. D. Francis, U. J. Dioriok, D. Liu, and M. J. Meaney, "Nongenomic Transmission Across Generations in Maternal Behavior and Stress Responses in the Rat," 1155–1158.

144   **neuroscientist Katharina Braun:** C. Helmeke, K. Seidel, G. Poeggel, T. W. Bredy, A. Abraham, and K. Braun, "Paternal Deprivation During Infancy Results in Dendrite- and Time-Specific Changes of Dendritic Development and Spine Formation in the Orbitofrontal Cortex of the Biparental Rodent *Octodon degus*," *Neuroscience* 163 (2009): 790–798.

145   **neuroscientist Catherine Belzung:** I. Yalcin, F. Aksu, and C. Belzung, "Effects of Desipramine and Tramadol in a Chronic Mild Stress Model in Mice Are Altered by Yohimbine but Not by Pindolol," *European Journal of Pharmacology* 514 (2005): 165–174.

146   **neuroscientist Allan Kalueff:** A. Denmark, D. Tien, K. Wong, A. Chung, J. Cachat, J. Goodspeed, C. Grimes, M. Elegante, C. Suciu, S. Elkhayat, B. Bartels, A. Jackson, M. Rosenberg, K. M. Chung, H. Badani, F. Kadri, S. Roy, J. Tan, S. Gaikwad, A. Stewart, I. Zapolsky, T. Gilder, and A. V. Kalueff, "The Effects of Chronic Social Defeat Stress on Mouse Self-Grooming Behavior and Its Patterning," *Behavioural Brain Research* 208 (2010): 553–559.

147   **our paternal California mice:** M. Bardi, C. L. Franssen, J. E. Hampton, E. A. Shea, A. P. Fanean, and K. G. Lambert, "Paternal Experience and Stress Responses in California Mice (*Peromyscus californicus*)," *Comparative Medicine* 61 (2011): 20–30.

148   ***Dalila effect*:** A. M. De Luca, "Environmental Enrichment: Does It Reduce Barbering in Mice?" *Animal Welfare Information Center*

Newsletter 8 (1997): 7–8; and A. F. Kalueff, A. Minasyan, T. Keisala, Z. H. Shah, and P. Tuohimaa, "Hair Barbering in Mice: Implications for Neurobehavioural Research," *Behavioural Processes* 71 (2006): 8–15; and J. R. Sarna, R. H. Dyck, and I. Q. Whishaw, "The Dalila Effect: C57BL6 Mice Barber Whiskers by Plucking," *Behavioural Brain Research* 108 (2000): 39–45; and F. A. R. Van den Broek, C. M. Omtzigt, and A. C. Beynen, "Whisker Trimming Behavior in A2G Mice Is Not Prevented by Offering Means of Withdrawal from It," *Laboratory Animals* 27 (1993): 270–272.

Chapter 8. The Final Rose Ceremony

154 **male rat greets the female rat by sniffing:** G. Bermant, "Copulation in Rats," *Psychology Today* 1 (1967): 52–60; and W. J. Jenkins and J. B. Becker, "Sex," in *The Behavior of the Laboratory Rat*, edited by I. Whishaw and B. Kolb, 307–320, New York: Oxford University Press, 2005.

155 **the male likes to sing:** M. K. McClintock, J. J. Anisko, and N. T. Adler, "Group Mating Among Norway Rats II—The Social Dynamics of Copulation: Competition, Cooperation, and Mate Choice," *Animal Behaivor* 30 (1982): 410–425; and N. R. White, R. Cagiano, A. U. Moises, and R. T. Barfield, "Changes in Mating Vocalization over the Ejaculatory Series in Rats (*Rattus norvegicus*)," *Journal of Comparative Psychology* 104 (1990): 255–262; and J. J. Anisko, S. F. Suer, M. K. McClintock, and N. T. Adler, "Relation Between 22-kHz Ultrasonic Signals and Sociosexual Behavior in Rats," *Journal of Comparative Physiology and Psychology* 92 (1978): 821–829.

156 **President Calvin Coolidge was touring a chicken farm:** T. Kealey, "The President, the First Lady, and a Rooster Who Just Couldn't Say No," *London Times*, August 8, 2005: www.timesonline.co.uk/tol/comment/columnists/guest_contributors/article552830.ece.

157 **when tested in the appropriate conditions:** W. J. Jenkins and J. B. Becker, "Sex," 307–320.

158 **a group of Italian neuroscientists wondered:** R. Cagiano, I. Bera, R. Sabatini, P. Flace, D. Vermesan, H. Vermesan, S. I. Dragulescu,

L. Bottalico, and L. Santacroce, "Effects on Rat Sexual Behaviour of Acute MDMA (Ecstasy) Alone or in Combination with Loud Music," *European Review for Medical and Pharmacological Sciences* 12 (2008): 285–293.

159   *ventromedial hypothalamic nucleus*: W. J. Jenkins and J. B. Becker, "Sex," 307–320.

159   **the reproductive hormone progesterone is delivered**: M. S. Erskine, "Solicitation Behavior in the Estrous Female Rat: A Review," *Hormones and Behavior* 23 (1989): 473–502; and M. M. McCarthy and J. B. Becker, "Neuroendorinology of Sexual Behavior in the Female," in *Behavioral Endocrinology*, edited by J. B. Becker, S. M. Breedlove, D. Crews, and M. M. McCarthy, 117–151, Cambridge, MA: MIT Press/Bradford Books, 2002.

160   **The brain's reward neurochemical (dopamine)**: F. J. Rivas and D. Mir, "Effects of Nucleus Accumbens Lesion on Female Rat Sexual Receptivity and Proceptivity in a Partner Preference Paradigm," *Behavioural Brain Research* 41 (1990): 239–249.

160   **study in the 1970s provided strong evidence**: B. J. Meyerson and L. Lindstrom, "Sexual Motivation in the Female Rat," *Acta physiologica Scandinavica* 389 (1973, Suppl.): 1–80.

161   **seemed to re-create Amsterdam's red-light district**: B. J. Everitt, "Sexual Motivation: A Neural and Behavioral Analysis of the Mechanisms Underlying Appetitive and Copulatory Responses of Male Rats," *Neuroscience and Biobehavioral Reviews* 14 (1990): 217–232.

161   **Natural increases of dopamine**: J. G. Pfaus, G. Damsma, G. G. Nomikos, D. G. Wenkstern, C. D. Blaha, A. G. Phillips, and H. C. Fibiger, "Sexual Behavior Enhances Central Dopamine Transmission in the Male Rat," *Brain Research* 530 (1990): 345–348.

162   **described as both stressful and rewarding**: B. Leuner, E. R. Glasper, and E. Gould, "Sexual Experience Promotes Adult Neurogenesis in the Hippocampus Despite an Initial Elevation in Stress Hormones," *PLoS ONE* 5 (2010): e11597, doi:101371/journal .pone.0011597.

165   **The more diverse an animal's MHC:** W. K. Potts, C. J. Manning, and E. K. Wakeland, "Mating Patterns in Seminatural Populations of Mice Influenced by MHC Genotype," *Nature* 352 (1991): 619–621.

166   **when the MHCs of couples:** R. Chaix, C. Cao, and P. Donnelly, "Is Mate Choice in Humans MHC-Dependent," *PLoS Genetics* 4 (2008): e10000184.

167   **Ohio State University neuroscientist:** S. L. Klein and R. J. Nelson, "Activation of the Immune-Endocrine System with Lipopolysaccharide Reduces Affiliative Behaviors in Voles," *Behavioral Neuroscience* 113 (1999): 1042–1048.

167   **In the 1990s, Leah Swongeur:** L. Swongeur, K. Lambert, and S. L. Klein, "Lipopolysaccharide-Activated Immune Function Influences Affiliative Behaviors and Mating Preferences in *Peromyscus californicus*," poster presented at the Society for Behavioral Neuroendocrinology meeting, Cincinnati, OH, 2003.

169   **Once prairie voles form a partner bond:** L. L. Getz and C. S. Carter, "Prairie-Vole Partnerships," *American Scientist* 84 (1996): 56–62.

169   **neuroscientist Sue Carter:** M. M. Cho, C. DeVries, J. R. Williams, and C. S. Carter, "The Effects of Oxytocin and Vasopressin on Partner Preferences in Male and Female Prairie Voles (*Microtus ochrogaster*)," *Behavioral Neuroscience* 113 (1999): 1071–1079.

169   **conducted with Karen Bales:** K. L. Bales, J. A. van Westerhuyzen, A. D. Lewis-Reese, N. D. Grotte, J. A. Lanter, and C. S. Carter, "Oxytocin Has Dose-Dependent Developmental Effects on Pair-Bonding and Alloparental Care in Female Prairie Voles." *Hormones and Behavior* 52 (2007): 274–279.

170   **if a vasopressin receptor gene:** S. Wrobel, "Coupling," *Emory/Health* (spring 2010), 2–4.

170   **the conservation of life's basic processes:** H. Walum, L. Westberg, S. Henningsson, J. M. Neiderhiser, D. Reissi, W. Igl, J. M. Ganiban, E. L. Spotts, N. L. Pedersen, E. Eriksson, and P. Lichtenstein, "Genetic Variation in the Vasopressin Receptor 1a Gene (AVPR1A)

Associates with Pair-Bonding Behavior in Humans," *Proceedings of the National Academy of Sciences U.S.A.* 37 (2008): 14153–14156.

170 **interesting oxytocin receptor patterns:** H. E. Ross and L. J. Young, "Oxytocin and the Neural Mechanisms Regulating Social Cognition and Affiliative Behavior," *Frontiers in Neuroendocrinology* 30 (2009): 534–547.

171 **dopamine plays a role in pair bonds:** L. A. McGraw and L. J. Young, "The Prairie Vole: An Emerging Model Organism for Understanding the Social Brain," *Trends in Neurosciences* 33 (2009): 103–109.

171 **simultaneously intrigued and scared:** P. Zak, "The Moral Molecule: The Oxytocin Cure," *Psychology Today* blog, November 10, 2008: www.psychologytoday.com/blog/the-moral-molecule/200811/the-oxytocin-cure.

171 **When pairs are separated by a partition:** T. G. Amstislavskaya and N. K. Popova, "Female-Induced Sexual Arousal in Male Mice and Rats: Behavioral and Testosterone Response," *Hormones and Behavior* 46 (2004): 544–550.

172 **James Pfaus at Concordia University:** N. Ismail, H. Gelez, I. Lachapelle, and J. G. Pfaus, "Pacing Conditions Contribute to the Conditioned Ejaculatory Preference for a Familiar Female in the Male Rat," *Physiology and Behavior* 96 (2009): 201–208.

Chapter 9. Family Values

173 **Quayle was clear about his position:** "Dan Quayle vs Murphy Brown," *Time*, June 1992: www.time.com/time/magazine/article/0,9171,975627,00.html.

173 **Social policy researcher Hilda Kahne:** H. Kahne, "Low-Wage Single-Mother Families in This Jobless Recovery: Can Improved Social Policies Help?" *Analyses of Social Issues and Public Policy* 4 (2004): 47–68.

175 **effects of prenatal stress on subsequent stress:** J. L. Humm, K. G. Lambert, and C. H. Kinsley, "Reductions in c-Fos Activity in the Medial Preoptic Area of Prenatally-Stressed Male Rats Following

Exposure to Sexually-Receptive Females," *Brain Research Bulletin* 37, no. 4 (1995): 363–368; and H. E. Jones, R. Rowe, B. Billack, C. Hancock, M. Ruscio, C. Gonzales, K. Lambert, and C. H. Kinsley, "Prenatal Stress Alters the Size of the Rostral Anterior Commissure in the Rats," *Brain Research Bulletin* 42 (1997): 341–346; and S. L. Klein, K. G. Lambert, D. Durr, T. Schaefer, and B. Waring, "The Influence of Environmental Enrichment and Sex on Predator-Stress Response in Rats," *Physiology and Behavior* 56 (1994): 291–297; and K. G. Lambert and C. H. Kinsley, "Sex Differences and Gonadal Hormones Influence Susceptibility to the Activity-Stress Paradigm," *Physiology and Behavior* 53, no. 6 (1993): 1085–1090; and K. G. Lambert, C. H. Kinsley, H. E. Jones, S. L. Klein, S. N. Peretti, and K. M. Stewart, "Prenatal Stress Attenuates Ulceration in the Activity-Stress Paradigm," *Physiology and Behavior* 57, no. 5 (1995): 989–994.

177   **a treasure trove of information about maternal brain:** R. S. Bridges and C. T. Grimm, "Reversal of Morphine Disruption of Maternal Behavior by Concurrent Treatment with the Opiate Antagonist Naloxone," *Science* 218 (1982): 166–168; and B. C. Nephew, E. M. Byrnes, and R. S. Bridges, "Vasopressin Mediates Enhanced Offspring Protection in Multiparous Rats," *Neuropharmacology* 58 (2010): 102–106; and M. Numan, "Medial Preoptic Area and Maternal Behavior in the Female Rat," *Journal of Comparative and Physiological Psychology* 87 (1974): 746–759; and M. Numan and T. R. Insel, *The Neurobiology of Parental Behavior,* New York: Springer, 2003; and M. Numan, J. S. Rosenblatt, and B. R. Komisaruk, "Medial Preoptic Area and Onset of Maternal Behavior in the Rat," *Journal of Comparative and Physiological Psychology* 91 (1977): 146–164; and P. E. Mann, C. H. Kinsley, P. M. Ronsheim, and R. S. Bridges, "Long-Term Effects of Parity on Opioid and Nonopioid Behavioral and Endocrine Responses," *Pharmacology, Biochemistry, and Behavior* 34 (1989): 83–88.

177   **Bridges and Numan completed their postdoctoral fellowships:** A. S. Fleming, M. Numan, and R. S. Bridges, "Father of Mothering: Jay S. Rosenblatt," *Hormones and Behavior* 55 (2009): 484–487.

181   **When the *Nature* paper was published:** C. H. Kinsley, L. Madonia, G. W. Gifford, K. Tureski, G. R. Griffin, C. Lowry, J. Williams,

J. Collins, H. McLearie, and K. G. Lambert, "Motherhood Improves Learning and Memory," *Nature* 402 (1999): 137.

182 **a longitudinal study to explore the long-term effects:** G. Love, N. Torrey, I. McNamara, M. Morgan, M. Banks, N. Wightman, E. R. Glasper, A. C. DeVries, C. H. Kinsley, and K. G. Lambert, "Maternal Experience Produces Long-Lasting Behavioral Modifications in the Rat," *Behavioral Neuroscience* 119 (2005): 1084–1096.

183 **older maternal rats were found to have less of a protein:** J. D. Gatewood, M. D. Morgan, M. Eaton, I. M. McNamara, L. F. Stevens, A. H. Macbeth, E. A. Meyer, L. M. Lomas, F. J. Kozub, K. G. Lambert, and C. H. Kinsley, "Motherhood Mitigates Aging-Related Decrements in Learning and Memory and Positively Affects Brain Aging in the Rat," *Brain Research Bulletin* 66, no. 2 (2005): 91–98.

185 **Fieldwork using DNA tests indicates:** D. O. Ribble, "The Monogamous Mating System of *Peromyscus californicus* as Revealed by DNA Fingerprinting," *Behavioral Ecology and Sociobiology* 29 (1991): 161–166.

185 **initial pilot studies suggest:** C. L. Franssen, C. L. Shea, J. E. Hampton, M. Bardi, J. Huber, M. M. Hyer, A. Rhone, R. A. Franssen, C. H. Kinsley, and K. G. Lambert, "Fatherhood Enhances Learning and Memory," poster presented at the International Behavioral Neuroscience Society annual meeting, Sardinia, Italy, 2010; and E. Glasper, K. Kaiss, C. Raffetto, and K. G. Lambert, "Reproductive Experience Alters Anxiety and Learning Ability in *Peromyscous californicus* Males and Females," poster presented at the International Behavioral Neuroscience Society annual meeting, Cancun, Mexico, 2001.

185 **mice dads were calmer:** M. Bardi, C. L. Franssen, J. E. Hampton, E. A. Shea, A. P. Fanean, and K. G. Lambert, "Paternal Experience and Stress Responses in California Mice (*Peromyscus californicus*)," 20–30.

187 **differences in parenting strategies:** K. G. Lambert, C. L. Franssen, M. Bardi, J. E. Hampton, L. Hainley, S. Karsner, E. B. Tu, M. M. Hyer, A. Crockett, A. Baranova, J. Ferguson, T. Ferguson, and C. H. Kinsley, "Characteristic and Distinct Neurobiological Patterns Differentiate Paternal Responsiveness in Two *Peromyscus* Species," unpublished manuscript.

192 **the research findings of James Swain:** J. E. Swain, R. Feldman, L. C. Mayes, and J. F. Leckman, "Parental Brain Emotion Circuits Vary with Gender, Correlate with Mood and Predict Behavior," poster presented at the International Behavioral Neuroscience Society annual meeting, Sardinia, Italy, 2010.

194 **poverty often also goes hand in hand with such conditions:** H. Kahne, "Low-Wage Single-Mother Families in This Jobless Recovery: Can Improved Social Policies Help?" *Analyses of Social Issues and Public Policy* 4 (2004): 47–68; and L. Mishel, J. Bernstein, and H. Boushey, *The State of Working America: 2002–2003*, Ithaca, NY: Cornell University Press, 2002.

194 **the extent of the effects of poverty:** A. S. Ivy, K. L. Brunson, C. Sandman, and T. Z. Baram, "Dysfunctional Nurturing Behavior in Rat Dams with Limited Access to Nesting Material: A Clinically Relevant Model for Early-Life Stress," *Neuroscience* 154 (2008): 1132–1142.

195 **Baram and her colleagues:** K. L. Brunson, E. Kramar, B. Lin, Y. Chen, L. L. Golgin, T. K. Yanagihara, G. Lunch, and T. Baram, "Mechanisms of Late-Onset Cognitive Decline After Early-Life Stress," *Journal of Neuroscience* 25 (2005): 9328–9338.

196 **neuroscientist Katharina Braun:** C. Helmeke, K. Seidel, G. Poeggel, T. W. Bredy, A. Abraham, and K. Braun, "Paternal Deprivation During Infancy Results in Dendrite- and Time-Specific Changes of Dendritic Development and Spine Formation in the Orbitofrontal Cortex of the Biparental Rodent *Octodon degus*," 790–798.

197 **wildlife biologist Ken Aplin:** P. D. MacLean, *The Triune Brain in Evolution: Role in Paleocerebral Functions*, New York: Plenum Press, 1990; and *Rat Attack*, Nova, 2009, video transcript: www.pbs.org/wgbh/nova/transcripts/3603_rats.html.

Chapter 10. Winning the Rat Race

202 **this neuronal dogma started to change:** I. Ortega-Perez, K. Murray, and P. M. Lledo, "The How and Why of Adult Neurogenesis," *Journal of Molecular Histology* 38 (2007): 555–562.

203 **four weeks is adequate for a mature neuron:** H. D. Schmidt and R. S. Duman, "The Role of Neurotrophic Factors in Adult Hippocampal Neurogenesis, Antidepressant Treatments and Animal Models of Depressive-Like Behavior," 391–418.

203 **Rusty Gage, a noted neurogenesis researcher:** P. S. Eridsson, E. Perfilieva, T. Bjork-Eridsson, A. Alborn, C. Nordborg, D. A. Peterson, and F. H. Gage, "Neurogenesis in the Adult Human Hippocampus," *Nature Medicine* 11 (1998): 1313–1317.

204 **Other areas, such as the neocortex:** E. Gould, "How Widespread Is Adult Neurogenesis in Mammals?" *Nature Reviews/Neuroscience* 8 (2007): 481–488.

206 **these studies suggest that voluntary running can produce:** P. J. Clark, R. A. Kohman, D. S. Miller, T. K. Bhattacharya, and E. H. Haferkamp, "Adult Hippocampal Neurogenesis and c-Fos Induction During Escalation of Voluntary Wheel Running in C567/6J Mice," *Behavioural Brain Research* 213 (2010): 246–252.

207 **Peter Clark and his colleagues:** Ibid.

207 **running-induced neurogenesis effect:** A. S. Naylor, C. Bull, M. K. L. Nilsson, C. Zhu, T. Bjork-Eriksson, P. S. Eridsson, K. Blomgren, and H. G. Kuhn, "Voluntary Running Rescues Adult Hippocampal Neurogenesis After Irradiation of the Young Mouse Brain," *Proceedings of the National Academy of Sciences U.S.A.* 105 (2008): 14632–14637.

208 **some Canadian rats revealed further interesting data:** J. M. Wojtowicz, M. L. Askew, and G. Winocur, "The Effects of Running and of Inhibiting Adult Neurogenesis on Learning and Memory in Rats," *European Journal of Neuroscience* 27 (2008): 1494–1502.

208 **A similar rescue effect was observed:** P. Larenetre, O. Leske, Z. Ma-Hogemeie, A. Haghikia, Z. Bichler, P. Wahle, and R. Heumann, "Exercise Can Rescue Recognition Memory Impairment in a Model with Reduced Adult Hippocampal Neurogenesis," *Frontiers in Behavioral Neuroscience* 3 (2010), doi:10.3389/neuro.08.034.2009.

209 **mice were forced to run on a motorized treadmill:** S. E. Kim, I. G. Ko, B. K. Kim, M. S. Shin, S. Cho, C. H. Kim, S. H. Kim, S. S. Baek,

E. K. Lee, and Y. S. Jee, "Treadmill Exercise Prevents Aging-Induced Failure of Memory Through an Increase in Neurogenesis and Suppression of Apoptosis in Rat Hippocampus," *Experimental Gerontology* 45 (2010): 357–365.

210 **Thirty days of wheel running:** R. A. Swain, A. B. Harris, E. C. Wiener, M. V. Dutka, H. D. Morris, B. E. Theien, S. Konda, K. Engberg, P. C. Lauterbur, and W. T. Greenough, "Prolonged Exercise Induces Angiogenesis and Increases Cerebral Blood Volume in Primary Motor Cortex of the Rat," *Neuroscience* 117 (2003): 1037–1046.

210 **ten days of wheel exposure enhanced blood flow:** K. Van der Borght, D. E. Kobor-Nyakas, K. Klauke, B. J. L. Eggen, C. Nyakas, E. A. Van der Zee, and P. Meerlo, "Physical Exercise Leads to Rapid Adaptations in Hippocampal Vasculature: Temporal Dynamics Relationship to Cell Proliferation and Neurogenesis," *Hippocampus* 19 (2009): 928–936.

210 **Thomas Hauser's team in Switzerland:** T. Hauser, F. Klaus, H. P. Lipp, and I. Amrein, "No Effect of Running and Laboratory Housing on Adult Hippocampal Neurogenesis in Wild Caught Long-Tailed Wood Mouse," *BMC Neuroscience* (2009), doi:10.1186/1471-2202-10-43.

211 **Liz Gould's team at Princeton:** A. M. Stranahan, D. Khalil, and E. Gould, "Social Isolation Delays the Positive Effects of Running on Adult Neurogenesis," *Nature Neuroscience* 9 (2006): 526–532.

212 **Karl Fernandes, a neuroscientist:** M. R. Bednarczyk, L. C. Hacker, S. Fortin-Nunez, A. Aumnt, R. Bergeron, and K. J. Fernandes, "Distinct Stages of Adult Hippocampal Neurogenesis Are Regulated by Running and the Running Environment," *Hippocampus* (2010), doi:10.1002/hipo.20831; and W. P. Pare, "Activity-Stress Ulcer in the Rat: Frequency and Chronicity," *Physiology and Behavior* 16 (1976): 699–704.

216 **the bacterium *Helicobacter pylori*:** B. J. Marshall and J. R. Warren, "Unidentified Curved Bacilli in the Stomach of Patients with Gastritis and Peptic Ulceration," *Lancet* 1 (1984): 1311–1315.

216 **My dissertation research was mildly interesting:** K. G. Lambert and L. J. Peacock, "Feeding Regime Affects Activity-Stress Ulcer Production," *Physiology and Behavior* 46, no. 4 (1989): 743–746.

216   **at Virginia Commonwealth University, Joe Porter:** K. G. Lambert and J. H. Porter, "Pimozide Mitigates Excessive Running in the Activity-Stress Paradigm," *Physiology and Behavior* 52 (1992): 299–304.

218   **I manipulated ambient temperature:** K. G. Lambert and L. Hanrahan, "The Effect of Ambient Temperature on the Activity-Stress Ulcer Paradigm," paper presented at the Southern Society for Philosophy and Psychology meeting, Louisville, KY, 1990.

218   **the running wheel activated a feral type of response:** J. G. Mather, "Wheel-Running Activity: A New Interpretation," *Mammal Review* 11 (2008): 41–51.

219   **uncovered several possible suspects:** K. G. Lambert, "The Activity-Stress Paradigm: Possible Mechanisms and Applications," *Journal of General Psychology* 120, no. 1 (1993): 21–32.

220   **Two neuroscientists, Jean Kant and Sally Anderson:** S. M. Anderson, G. A. Saviolakis, R. A. Bauman, K. Y. Chu, S. Ghosh, and G. J. Kant, "Effects of Chronic Stress on Food Acquisition, Plasma Hormones, and the Estrous Cycle of Female Rats," *Physiology and Behavior* 60 (1996): 325–329; and G. J. Kant, R. H. Pastel, R. A. Bauman, G. R. Meininger, K. R. Maughan, T. N. Robinson, W. L. Wright, and P. S. Covington, "Effects of Chronic Stress on Sleep in Rats," *Physiology and Behavior* 57 (1995): 359–365.

222   **the therapeutic effects of running:** J. M. Halperin and D. M. Healey, "The Influences of Environmental Enrichment, Cognitive Enhancement, and Physical Exercise on Brain Development: Can We Alter the Developmental Trajectory of ADHD?" *Neuroscience and Biobehavioral Reviews* (2010), doi:10.1016/j.neubiorev2010.07.006; and N. F. Ho, S. P. Han, and G. S. Dawe, "Effect of Voluntary Running on Adult Hippocampal Neurogenesis in Cholinergic Lesioned Mice," *BMC Neuroscience* (2009), doi:10.1186/1471-2202-10-57; and R. Hoveida, H. Alaei, S. Oryan, K. Parivar, and P. Reisi, "Treadmill Running Improves Spatial Memory in an Animal Model of Alzheimer's Disease," *Behavioural Brain Research* (2010), doi:10.1016/j.bbr.2010.08.003.

222   **Merzenich has developed behavioral training strategies:** M. Fisher, C. Holland, M. M. Merzenich, and S. Vinogradov, "Using

Neuroplasticity-Based Auditory Training to Improve Verbal Memory in Schizophrenia," *American Journal of Psychiatry* 166 (2009): 805–811.

223 **Posit Science, has developed engaging interactive software:** Scientific Learning, "Dr. Michael M. Merzenich": www.scilearn.com/our-approach/our-scientists/merzenich.

223 **"It's medicine," Merzenich declared:** Ginger Campbell, *Brain Science Podcast*, February 13, 2009: www.brainsciencepodcast.com/bsp/michael-merzenich-talks-about-neuroplasticity-bsp-54.html.

223 **effectiveness of antidepressants such as fluoxetine (Prozac):** I. Kirsch, *The Emperor's New Drugs: Exploding the Antidepressant Myth*, New York: Basic Books, 2010.

223 **neuroscientist Henriette van Praag:** M. W. Marlatt, P. J. Lucassen, and H. van Praag, "Comparison of Neurogenic Effects of Fluoxetine, Duloxetine and Running in Mice," *Brain Research* (2010): 93–99.

224 **As I expressed in a recent book:** K. Lambert, *Lifting Depression*.

Chapter 11. Weapons of Mass Destruction

229 **behavioral scientist John Calhoun:** J. B. Calhoun, "Population Density and Social Pathology," *Scientific American* 306 (1962): 139–148; E. Ramsden, "The Urban Animal: Population Density and Social Pathology in Rodents and Humans," *Bulletin of the World Health Organization* 87 (2009): 82; and E. Ramsden and J. Adams, "Escaping the Laboratory: The Rodent Experiments of John B. Calhoun and Their Cultural Influence [Abstract]," *Journal of Social History* 42, no. 3 (2009): http://chnm.gmu.edu/jsh/abstracts.php?volume=42&issue=3#963.

230 **rat crowding can be found the Karni Mata Temple:** S. Guynup and N. Ruggia, "Rats Rule at Indian Temple," *National Geographic*, June 29, 2004: http://news.nationalgeographic.com/news/2004/06/0628_040628_tvrats.html.

231 **S. A. Barnett described the interaction:** S. A. Barnett, *The Rat*.

232 **behavioral neuroscientists Klaus Miczek and Sietse F. de Boer:** K. A. Miczek and S. de Boer, "Aggressive, Defensive, and Submissive

Behavior," in *The Behavior of the Laboratory Rat*, edited by I. Whishaw and B. Kolb, 344–352, New York: Oxford University Press, 2005.

234 **the fighting response isn't nearly as interesting:** K. L. Huhman, M. G. Solomon, M. Janicki, A. C. Harmon, S. M. Lin, J. E. Israel, and A. M. Jasnow, "Conditioned Defeat in Male and Female Syrian Hamsters," *Hormones and Behavior* (2003): 293–299.

235 **human adolescents who are not only victims:** R. Kaltiala-Heino, M. Rimpela, M. Martunen, A. Rimpela, and P. Rantanen, "Bullying, Depression, and Suicidal Ideation in Finnish Adolescents: School Survey," *British Medical Journal* 319 (1999): 348–351; and A. H. Luukkonen, P. Rasanen, H. Hakko, and K. Riala, "Bullying Behavior Is Related to Suicide Attempts but Not to Self-Mutilation Among Psychiatric Inpatient Adolescents," *Psychopathology* 42 (2009): 131–138; and I. Rivers and N. Noret, "Participant Roles in Bullying Behavior and Their Association with Thoughts of Ending One's Life," *Crisis* 31 (2010): 143–148.

236 **There is, however, a glimmer of hope:** S. A. Erath, K. S. Flanagan, K. L. Bierman, and K. M. Tu, "Friendships Moderate Psychosocial Maladjustment in Socially Anxious Early Adolescents," *Journal of Applied Developmental Psychology* 31 (2010): 15–26.

236 **Eberhard Fuchs and his colleagues:** B. Czeh, N. Abumaria, R. Rygula, and E. Fuchs, "Quantitative Changes in Hippocampal Microvasculature of Chronically Stressed Rats: No Effect of Fluoxetine Treatment," *Hippocampus* 20 (2010): 174–185; and C. J. Herzog, B. Czeh, S. Corbach, W. Wuttke, O. Schulte-Herbruggen, R. Hellweg, G. Flugge, and E. Fuchs, "Chronic Social Instability Stress in Female Rats: A Potential Animal Model for Female Depression," *Neuroscience* 19 (2009): 982–992.

237 **psychological warfare, accompanied by a deluge of stress hormones:** J. Cole, A. W. Toga, C. Hojatkashani, P. Thompson, S. G. Costafreda, A. J. Cleare, S. C. R. Williams, E. T. Bullmore, J. L. Scott, M. T. Mitterschiffthaler, N. D. Walsh, C. Donaldson, M. Mirza, A. Marquand, C. Nosarti, P. McGuffin, and C. H. Y. Fu, "Subregional Hippocampal Deformations in Major Depressive Disorder," *Journal of Affective Disorders* 126 (2010): 272–277.

238 **social philanthropist, Bart Weetjens:** Cassandra Herrman, director, *Tanzania: Hero Rats,* video: www.pbs.org/frontlineworld/
stories/tanzania605/video_index.html; and E. McLaughlin, "Giant
Rats Put Nose to Work on Africa's Land Mine Epidemic," CNN
World, 2010: www.cnn.com/2010/WORLD/africa/09/07/herorats
.detect.landmines/index.html.

241 **entertaining video tribute to them:** Parry Gripp, *Hero Rats,* video:
www.youtube.com/watch?v=NR5VAvlz200.

Chapter 12. Keeping It Real: The Most Important Lesson of All

243 **a male Norway rat wandered into a trap:** J. C. Rusell, D. R. Towns,
S. H. Anderson, and M. N. Clout, "Intercepting the First Rat
Ashore," *Nature* 437 (2005): 1107.

244 **best-practice devices in rodent detection:** Ibid, quotation from
"Supplementary Methods": www.stat.auckland.ac.nz/~jrussell/files/
papers/4371107a-s1.doc.

247 **animal behaviorist Manuel Berdoy:** M. Berdoy, *The Laboratory Rat:
A Natural History,* film, Oxford University, 2002: www.RATLIFE.org.

251 **Sabra Klein, is in the same department:** J. D. Easterbrook, J. B.
Kaplan, N. B. Vanasco, W. K. Reeves, R. H. Purcell, M. Y. Kosoy, G.
E. Glass, J. Watson, and S. L. Klein, "A Survey of Zoonotic Patho-
gens Carried by Norway Rats in Baltimore, Maryland, USA," *Epide-
miological Infections* 135 (2007): 1192–1199; and J. D. Easterbrook, J.
G. Kaplan, G. E. Glass, M. V. Pletnikov, and S. L. Klein, "Elevated
Testosterone and Reduced 5-HIAA Concentrations Are Associated
with Wounding and Hantavirus Infection in Male Norway Rats,"
Hormones and Behavior 52 (2007): 474–481; and L. C. Gardner-Santana,
D. E. Norris, C. M. I. Fornadel, E. R. Hinson, S. L. Klein, and G. E.
Glass, "Commensal Ecology, Urban Landscapes, and Their Influ-
ence on the Genetic Characteristics of City-Dwelling Norway Rats
(*Rattus norvegicus*)," *Molecular Ecology* 18 (2009): 2766–2778; and C.
Landers, "Rats: They're Everywhere, They Spread Disease—And
They're Gaining on Us," *Baltimore City Paper,* October 10, 2008:
www2.citypaper.com/news/story.asp?id=12787.

256   documentary *An Inconvenient Truth*: S. Lovgren, "Al Gore's 'Inconvenient Truth' Movie: Fact or Hype?" *National Geographic News*, May 25, 2006: http://news.nationalgeographic.com/news/2006/05/060524-global-warming.html.

256   In my book *Lifting Depression*: K. G. Lambert, *Lifting Depression*.

257   biological paleoanthropologist John Hawks's work: C. Q. Choi, "Humans Still Evolving as Our Brains Shrink," *LiveScience*, November 13, 2009: www.livescience.com/history/091113-origins-evolving.html.

259   In *Anatomy of an Epidemic*: R. Whitaker, *Anatomy of an Epidemic*, New York: Crown, 2010.

259   there is no clear evidence that drugs such as Prozac: J. E. Castro, E. Varea, C. Marquez, M. I. Cordero, G. Poirier, and C. Sandi, "Role of the Amygdala in Antidepressant Effects on Hippocampal Cell Proliferation and Survival and on Depression-Like Behavior in the Rat," *PLoS ONE* 5, no. 1 (2010): e8618, doi:10.1371/journal.pone.0008618; and L. Dazzi, E. Seu, G. Cherchi, and G. Biggio, "Chronic Administration of the SSRI Fluvoxamine Markedly and Selectively Reduces the Sensitivity of Cortical Serotonergic Neurons to Footshock Stress," *European Neuropsychopharmacology* 15 (2005): 283–290; and A. M. Gardier, E. Lepoul, J. H. Trouvin, E. Chanut, M. C. Dessalies, and C. Jacquot, "Changes in Dopamine Metabolism in Rat Forebrain Regions After Cessation of Long-Term Fluoxetine Treatment: Relationship with Brain Concentrations of Fluoxetine and Norfluoxetine," *Life Sciences* 54 (1994): PL51–56; and G. Griebel, D. C. Blanchard, R. S. Agnes, and R. J. Blanchard, "Differential Modulation of Antipredator Defensive Behavior in Swiss-Webster Mice Following Acute or Chronic Administration of Imipramine and Fluoxetine," *Psychopharmacology* 120 (1995): 37–66; and M. E. Page and E. D. Abercrombie, "An Analysis of the Effects of Acute and Chronic Fluoxetine on Extracellular Norepinephrine in the Rat Hippocampus During Stress," *Neuropsychopharmacology* 16 (1997): 419–425; and G. T. Taylor, S. Farr, K. Klinga, and J. Weiss, "Chronic Fluoxetine Suppresses Circulating Estrogen and the

Enhanced Spatial Learning of Estrogen-Treated Ovariectomized Rats," *Psychoneuroendocrinology* 29 (2004): 1241–1249.

260   **at least 10 percent of our population taking antidepressants:** M. Olfson and S. C. Marcus, "National Patterns in Antidepressant Medication Treatment," *Archives of General Psychiatry* 66 (2009): 848–856.

INDEX

|||||||||||||||||||||||||||||||||||||||||||||||||||||||||||||||||||||||||||||||||||||

ABOUT THE AUTHOR

||||||||||||||||||||||||||||||||||||||||||||||||||||||||||||||||||||||||||||||||||||

Deb Silbert Photography

Kelly Lambert is the Macon and Joan Brock Professor and Chair of Psychology at Randolph-Macon College. In addition to teaching psychology and neuroscience courses, she maintains a behavioral neuroscience laboratory where she and her students investigate the plasticity of the mammalian brain. Currently serving as the president of the International Behavioral Neuroscience Society, she has published research articles in journals such as *Nature, Scientific American*, and *Behavioral Neuroscience*. She was named the 2008 Virginia Professor of the Year and, in 2001, received the State Council of Higher Education for Virginia Outstanding Faculty Award. Lambert's books include *Clinical Neuroscience: Psychopathology and the Brain* (with coauthor Craig Kinsley, Oxford University Press, 2010) and *Lifting Depression: A Neuroscientist's Hands-On Approach to Activating Your Brain's Healing Power* (Basic Books, 2008). She lives in Mechanicsville, Virginia, with her husband, Gary, an industrial-organizational psychologist; her two daughters, Lara and Skylar; and several pets (an intelligent dog named Golgi, a suspicious cat named Gracie, and two volcanic-ash-loving chinchillas, Sophie and Sadie).